BAYESIAN NETWORKS

Edited by **Wichian Premchaiswadi**

Bayesian Networks

Edited by Wichian Premchaiswadi

Published by InTech

Janeza Trdine 9, 51000 Rijeka, Croatia

Publishing Process Manager Ivona Lovric

Technical Editor Teodora Smiljanic

Cover Designer InTech Design Team

Additional hard copies can be obtained from orders@intechopen.com

Bayesian Networks, Edited by Wichian Premchaiswadi
p. cm.
ISBN 978-953-51-0556-5

For EU product safety concerns:
IN TECH d.o.o., Prolaz Marije Krucifikse Kozulić 3, 51000 Rijeka, Croatia,
info@intechopen.com or visit our website at intechopen.com.

Meet the editor

Dr. Wichian Premchaiswadi received the M.Eng. and Ph.D. degrees in Electrical Engineering from Waseda University Japan. He is currently the Dean of the Graduate School of Information Technology in Business and Assistant President of Siam University, Thailand. He received the academic award for the best research conducted in 2008 from the Association of Private Higher Education Institution of Thailand. At present, he also works as the Secretary General of The Computer Association of Thailand under The Royal Patronage of His Majesty the King. His research interests include robust methods for autonomous computer vision, database systems, artificial intelligence, collaborative computing, human-computer interaction, decision support systems, group decision support systems, group decision making, parallel processing, image processing, nonparametric analysis, real-time vision systems, data mining, and pattern recognition.

Contents

Preface

Over the last decade, a Bayesian network has become a popular representation for encoding uncertain expert knowledge in expert systems. A Bayesian network is a graphical model for probabilistic relationships among a set of variables. It is a graphical model that encodes probabilistic relationships among variables of interest. When used in conjunction with statistical techniques, the graphical model has several advantages for data modeling. So what do Bayesian networks and Bayesian methods have to offer? There are at least four benefits described in the following.

One, Bayesian networks can readily handle incomplete data sets. For example, consider a classification or regression problem where two of the explanatory or input variables are strongly anti-correlated. This correlation is not a problem for standard supervised learning techniques, provided all inputs are measured in every case. When one of the inputs is not observed, however, many models will produce an inaccurate prediction, because they do not encode the correlation between the input variables. Bayesian networks offer a natural way to encode such dependencies.

Two, Bayesian networks allow one to learn about causal relationships. Learning about causal relationships is important for at least two reasons. The process is useful when we are trying to gain an understanding about a problem domain, for example, during exploratory data analysis. In addition, knowledge of causal relationships allows us to make predictions in the presence of interventions. For example, a marketing analyst may want to know whether or not it is worthwhile to increase exposure of a particular advertisement in order to increase the sales of a product. To answer this question, the analyst can determine whether or not the advertisement is a cause for increased sales, and to what degree. The use of Bayesian networks helps to answer such questions even when no experimental data about the effects of increased exposure is available.

Three, Bayesian networks, in conjunction with Bayesian statistical techniques, facilitate the combination of domain knowledge and data. Anyone who has performed a real-world modeling task knows the importance of prior or domain knowledge, especially when data is scarce or expensive to obtain. The fact that some commercial systems (i.e., expert systems) can be built from prior knowledge alone is a testament to the power of prior knowledge. Bayesian networks have a causal semantics that makes the encoding of causal prior knowledge particularly straightforward. In addition,

Bayesian networks encode the strength of causal relationships with probabilities. Consequently, prior knowledge and data can be combined with well-studied techniques from Bayesian statistics.

Four, Bayesian methods in conjunction with Bayesian networks and other types of models offer an efficient and principled approach for avoiding the over fitting of data.

This book deals with the theory and algorithms for learning and probabilistic inference in Bayesian networks. This book also provides selected applications of Bayesian networks in several fields, including adaptive risk management, operational risk analysis, rangeland management and resiliency and interdependency of critical urban infrastructure. The book chapters are original manuscripts written by experienced researchers that have made significant contributions to the field of Bayesian networks. Although all chapters are self- contained, the reader should be familiar with texts using mathematical and statistical language to gain full benefit from the book. I am convinced that this book will be a very useful tool for anyone who is concerned with modelling systems containing causality with inherent uncertainty. I believe that readers will not only find the technical aspects for using and implementing Bayesian networks to solve their problem, but will also discover new approaches for how their current research and work can benefit from one of the major tools of the 21st century.

Wichian Premchaiswadi
Graduate School of Information Technology in Business, Siam University,
Thailand

1

Making a Predictive Diagnostic Model for Rangeland Management by Implementing a State and Transition Model Within a Bayesian Belief Network (Case Study: Ghom- Iran)

Hossein Bashari

Department of Natural Resources, Isfahan University of Technology, Isfahan, Iran

1. Introduction

Rangelands are landscapes in which the native vegetation (climax or natural potential) is predominantly grasses, grass-like plants, forbs, or shrubs. They are defined as land where vegetation management is accomplished mainly through the manipulation of grazing and include land that is re-vegetated naturally or artificially (SRM, 1989). The world's rangelands are grazed because they do not have the capacity to be cultivated. They are globally significant semi-natural landscapes that have been used for many purposes including grazing, apiculture, hunting, mining and tourism. The degradation of rangelands, which cover more than 47 % of the globe (332 million hectares) (Tueller, 1988), has been reported in all parts of the world. For instance, more than 70% of rangelands in Africa, Asia and America and about 54% in Australia are to some degree degraded. Better understanding of the ecological processes in rangelands and of the products they provide are required to effectively maintain and manage this valuable resource.

Rangelands are highly complex adaptive socio-ecological systems with complicated interactions between the rangelands, livestock and humans (Gross et al., 2003; Gross et al., 2006). Leohle (2004) categorised the sources of ecological complexity, which are notable in rangelands, into six groups: spatial, temporal, structural, process, behavioural and geometric. Interactions between these components in a broad range of spatial and temporal scales are among the main reasons for their complexity. A lack of understanding in any part leads to an inability to identify the best policies and strategies for management (Walker & Janseen, 2002). Misunderstanding of these interactions by humans is responsible for a worldwide deterioration in rangeland ecosystems. The inherent complexity of ecological parameters and uncertain social and economic effects significantly adds to the difficulties of developing a sound understanding of rangelands. In addition, there may also be conflicts in the multiple objectives of rangeland use and management (e.g. production and conservation). Anti-degradation programs fail if they do not consider the interactions between the ecological, social and economic parameters within rangeland ecosystems.

The lack of availability of scientific knowledge (research results and experiences) at the time of decision-making by the different stakeholders and policy makers is one of the reasons for the failure of rangeland management programs. This knowledge is scattered over a wide range of resources and is not easily accessible even for scientists. In addition, the lack of integrating scientific knowledge, with landholders' knowledge and the slow response to the uptake of new knowledge by land managers hinders the success of management programs (Bosch et al., 2003).

It could be argued that most knowledge is available, but the formats in which rangeland managers would require such knowledge is often not accessible in an appropriate form. Translation of knowledge into practical applications is a prerequisite if this knowledge is to be used in management programs (Provenza, 1991).

1.1 Decision support tools in range management

Many simulation models have been developed by researchers for the purpose of predicting the outcomes of rangeland management decisions. These models help to:

- Organise and structure different sources of knowledge about rangeland systems;
- Identify and focus on the knowledge gaps;
- Promote a multidisciplinary approach to rangeland management;
- Provide an efficient means of capturing the complex dynamics of rangeland systems (Carlson et al., 1993).

There are many Decision Support Tools (DSTs) relevant to rangeland management that are based on simulation models (Carlson et al., 1993). Some of these DSTs have been specified for a single purpose or are appropriate for limited objectives or areas of application, while some have wider application. However, most models have been developed as research tools which require large data inputs. A good understanding of data requirements is needed for all models to assess their application and to evaluate their appropriateness and output value (National Land & Water Audit, 2004). This makes them inaccessible to most land managers.

An additional difficulty is the fact that uncertainty in the prediction of management outcomes is not accommodated in these DSTs. Uncertainty exists when there is more than one outcome, consistent with the expectations (Pielke, 2001, 2003). Decision-makers are interested in quantifying and reducing uncertainty. The degree of confidence in model predictions is therefore an important aspect to be included in the design of useful DSTs. Finally, it is the decision-makers task to understand and use the DSTs. It is therefore important that they are involved in developing the tools. Using the end-users' experiential knowledge could play a vital role in ensuring credibility and increasing the adoption of a DST.

1.2 Adaptive management

Adaptive management has become an important approach to cope with uncertainty, imperfect knowledge and complex systems. In this approach, outcomes of management are continuously used to modify or adapt management (Sabine et al., 2004; Morghan et al., 2006). This is particularly important for rangelands where the outcomes of management are often unknown or difficult to predict. Adaptive management "structures a system in which monitoring iteratively improves the knowledge base and helps refine management plans" (Ringold et al., 1996, P.745). However, a framework is needed that allows for this knowledge to be updated and ensures its accessibility for future decision-making. Such a framework must be able to predict the probable outcomes of rangeland management decisions based on the existing knowledge of vegetation dynamics. Such a framework should also accommodate the uncertainty associated with these predictions.

1.3 Framework for adaptive management

The State and Transition Model (STM) provides a simple framework for integrating knowledge about vegetation dynamics and the possible responses of vegetation to management actions and environmental events. Both qualitative and quantitative knowledge could be accommodated in an STM, and it has the potential for organising and updating new knowledge that becomes available through monitoring (Vayssieres & Plant, 1998). The STM is also ideal for improving communication between rangeland scientists, end users and policy makers. Using this model as a framework enables landholders to focus on opportunities (e.g. transition to productive states) and avoid hazards (e.g. transitions to degraded states where the reverse transitions are problematical since they will be too costly to reverse or not be practical in a normal management time scale) (Westoby et al., 1989; Brisk et al., 2005).

Many scientists have used these concepts for developing STMs for various rangelands following their introduction by Westoby et al. (1989), who introduced this model based on non-equilibrium ecology (Friedel, 1991; Laycock, 1991; Hall et al., 1994; Allen-Diaz & Bartolome, 1998; Phelps & Bosch, 2002; Knapp et al. 2011). Typically most of the STMs presented so far are simple flowcharts with a catalogue of states and transitions. These models are traditionally descriptive and are unable to be used as predictive models. Also, most of the models produced so far are characterised by a lack of practical application and simply provide "proof- of concept examples" (Vayssieres & Plant, 1998). However, they handle poorly the uncertainty associated with cause and effect.

Bayesian Belief Networks (BBNs) (Jensen, 1996) provide a tool that can help solve this problem. They allow for the construction of cause and effect models, and relate variables using conditional probabilities. This allows for uncertainty to be explicitly incorporated into models. BBNs can also be used to perform sensitivity and scenario analysis, allowing managers to predict the probable outcomes of management actions or identify those management actions that are most likely lead to desirable outcomes. An added benefit of BBNs is that they can be easily updated using the results of monitoring. These monitoring results can be used to update the probability relationships over time, allowing the outcomes of previously implemented management decisions to modify model predictions. Therefore, BBNs provide a mechanism for supporting adaptive management.

This chapter aimed to demonstrate how a STM can be converted into a user-friendly management decision support tool. This includes several steps including (a) converting the State and Transition diagram into a BBN influence diagram, (b) determining probabilities for the BBN model through literature studies and the knowledge of scientists that are familiar with the vegetation dynamics of the study area and finally (c) testing model behaviour by sensitivity and scenario analysis. A STM for the Steppe zone of Qom- Iran was used as an example to demonstrate the process.

2. Case study in Iran

2.1 Study area

The study area is located in Ghom rangelands, 130 km from Tehran, the capital city of the Islamic Republic of Iran. This area is surrounded by central Iranian desert and has an arid climate. The Steppe zone of Ghom has an annual rainfall of between 100 to 230 mm, which is highly variable both within and between seasons. Most precipitation occurs in winter with the dry season occurring for 4 to 6 months over summer. A significant portion of the limited precipitation is lost as run-off and then evaporation.

The vegetation in the Steppe zone of Ghom is sparse with evenly distributed individual dwarf shrubs and/or bunchgrasses. The perennial cover can vary from 1 to 35%, while the spaces between perennials remain bare or briefly covered by Therophytes after rainfall events. The most common life-form is shrubby species (browse species) and subshrubs (dwarf shrubs). The contribution of the subshrubs is about 40% of the perennial species, while about 30% of those are shrubby species. The most frequently occurring species is *Artemisia sieberi*. Woody plants that grow with *Artemisia sierberi* are *Dendrostellera lessertii, Ephedra sp, Astragalus sp, Achillea sp, scariola orientalis, Acantholimon sp, Acanthophylum sp,* and *Stachys inflate*. *Stipa hohenackeriana* is the most abundant perennial grass but it has disappeared from some areas. *Stipagrostis plumose* is another dominant perennial grass, however, is only found on light soils and never on heavy or saline soils

The Bureau of Rangelands has developed several strategies to enhance rangeland condition, including de-stocking, water harvesting and transplanting of palatable shrubs. The challenge is when and where to implement these strategies to obtain the best result. The effect of these strategies on the dynamics of the vegetation is also uncertain. The unavailability of an appropriate DST hinders the selection of appropriate management strategies.

2.2 Creating a State and Transition Model (STM)

The iterative model development process was used to construct an STM for the Steppe zone of Ghom Iran. This process utilised multiple information sources to identify possible vegetation states and transitions.

There were no published STMs for this area, so the process utilised multiple information sources to identify possible vegetation states and transitions. First, the limited literature available was used to draft a catalogue of states and transitions. Then it was refined through discussion with scientists familiar with the vegetation dynamics of the Ghom area. Vegetation states were defined using vegetation composition and soil erosion status. The favourability of each state was explained from an animal productivity and soil stability point of view.

Figure 1 shows the STM developed for the Steppe zone of Ghom. The STM consists of 7 vegetation states and 15 transitions (see Table 1 & Table 2).

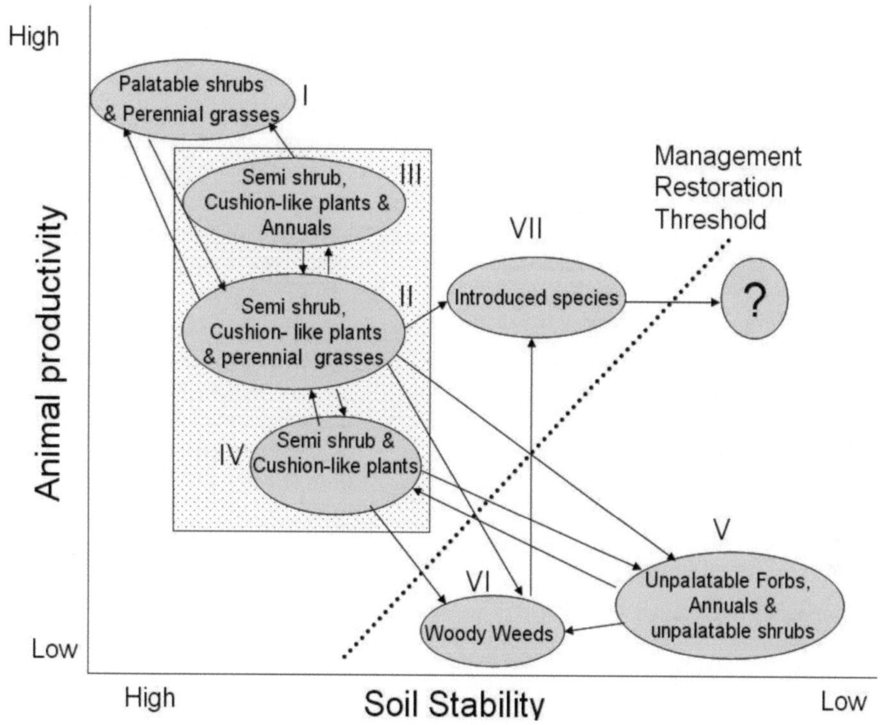

Fig. 1. STM for Steppe zone of Ghom, Iran based on the experiential knowledge of Iranian researchers.

Number of states	State of vegetation	Species Composition	Maximum frequency (%)	Some Ecological Information
I	Palatable shrubs & perennial grasses	*Artemisia sieberi*	15	This state has a high grazing value and generally has high litter, and projected cover of 30%. Pasture yield in this state is high; erosion level is low because it is dominated by shrubs and perennial grasses; thus, soil stability is high.
		Buffunia mucrocarpa	10	
		Pteropyron sp	5	
		Salsola tomentosa	5	
		Andrachne sp	5	
		Ajuga sp	5	
		Kochia sp	5	
		Stipa hohenackeriana	20	
		Other	30	
II	Semi shrub, cushion-like plants & perennial grasses	*Artemisia sieberi*	60	Erosion is low to moderate. This state represents the overall condition in the Steppe zone.
		Cushion-like plants	10	
		Stipa hohenackeriana	5	
		Other	25	

Table 1. Continued

Number of states	State of vegetation	Species Composition	Maximum frequency (%)	Some Ecological Information
III	Semi-shrub, cushion-like plants and annuals	*Artemisia sieberi* Cushion-like plants *Stipa hohenackeriana* Annuals	60 10 5 25	Its composition is the same as state II, but is highly preferred by sheep and goats due to abundant foliage of annuals. It can have up to 30% cover of annual grasses and forbs.
IV	Semi-shrub & cushion-like plants	*Artemisia sieberi* Cushion-like plants *Noaea mucronata* *Stipa hohenackeriana* Others	50 30 5 1 14	In this state, palatable shrubs such as *Salsola tomentosa* have disappeared and frequency of tall grass species such as *Stipa hohenackeriana* decreases dramatically. Erosion is high.
V	Unpalatable forbs, annuals and unpalatable shrubs	*Peganum harmala* *Launaea acanthodes* *Euphorbia spp* Cushion-like plants *Artemisia sieberi* *Noaea mucronata* *Scariola orientalis* Annual grass Annual forbs Other	5 5 5 5 20 10 10 5 5 30	This represents the most degraded state, there are no perennial grasses in this state and also *Artemisia sieberi* has a low frequency. The percentage cover is less than 10% and erosion is high.
VI	Woody weeds	*Reseda sp* *Hulthemia persica*	10 90	Some species (*Reseda sp* and *Hulthemia persica)* have infested these areas that were formerly cultivated. A highly stable state with lowest value for grazing.
VII	Introduced species	*Atriplex spp* *Artemisia sieberi* Cushion-like plants *Stipa hohenackeriana* Others	50 20 10 5 15	This state has two stratifications. *Atriplex spp* constitutes the upper while various other species are located in the lower level of the vegetation structure. The total percentage cover is low and the frequency of species such as *Artemisia sieberi* declines to the that of state V.

Table 1. Catalogue of vegetation states for Steppe zone of Ghom.

Transition number & name		Main causes	Probability	Time frame (years)
1	I, II	Grazing pressure (Moderate), Selective grazing (High), Early grazing	High	5-10
2	II, I	Destock, Wet season in time period (Frequent), Seed and plant of palatable shrub available	High	>10
3	II, III	Wet season in time period (Frequent)	High	1-2
4	II, IV	Grazing pressure (High), Drought, Soil compaction (High)	High	3-10
5	II, V	Grazing pressure (Very high), Drought (Frequent), Soil compaction (High)	High	5-20
6	II, VI	Ploughing	High	2-5
7	II, VII	Transplanting Seedling of *Atriplex spp*, Wet season in time period (Frequent),Irrigation of seedlings in initial years	High	3-5
8	III, I	Destock, Wet season in time period (Frequent), Seed and plant sources	Moderate	>10
9	III, II	Wet season in time period (Infrequent)	High	1-2
10	IV, II	Grazing pressure (low), Wet season in time period (Frequent), Seeds and plant sources decrease	High	5-10
11	IV, V	Grazing pressure (High), Drought (Frequent), Soil compaction (High)	High	5-10
12	IV, VI	Ploughing	High	2-5
13	V, IV	Grazing pressure (low), Wet season in time period (Frequent), Seed and plant sources	Low to moderate	2-5
14	V, VI	Ploughing	High	2-5
15	VI, VII	Erasing Woody weeds, Plantation of *Atriplex spp*, Wet season in time period (Frequent)	Moderate	3-5

Table 2. Catalogue of transitions for the Steppe zone of Ghom.

2.3 Creating a BBN for modelling vegetation change

Figure 2 outlines the main steps used in this study to build a DST for rangeland management. The STM for Stepp zone of Ghom (outlined above) was the starting point for model development. From the STM, an influence diagram was built to show the possible transitions and the factors influencing each transition. Next, the influence diagram was populated with probabilities to produce a predictive model, and finally the behaviour of the model was tested using scenario and sensitivity analysis.

2.4 Building an influence diagram

An influence diagram is simply the graphical component of a BBN. From the STM, an influence diagram was constructed to show the possible transitions and the factors influencing each. The framework contains a node representing possible starting vegetation states, nodes representing possible transitions from each of these states to other states, nodes representing the main factors influencing these transitions and their sub-factors , and time frame of possible changes.

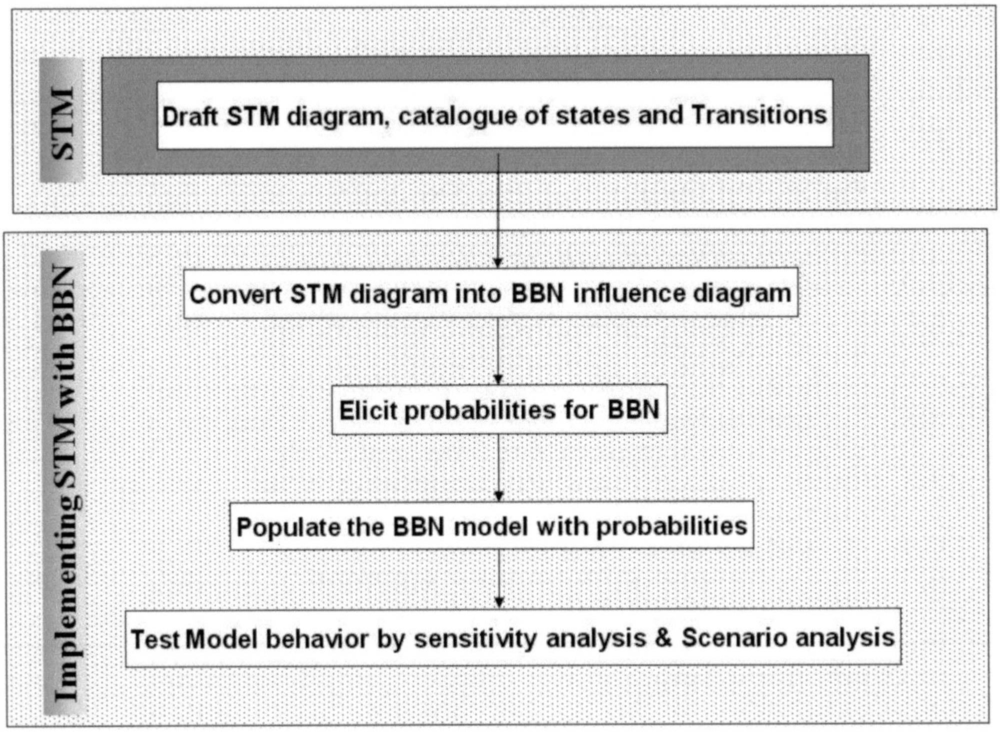

Fig. 2. Steps used to build a decision support tool.

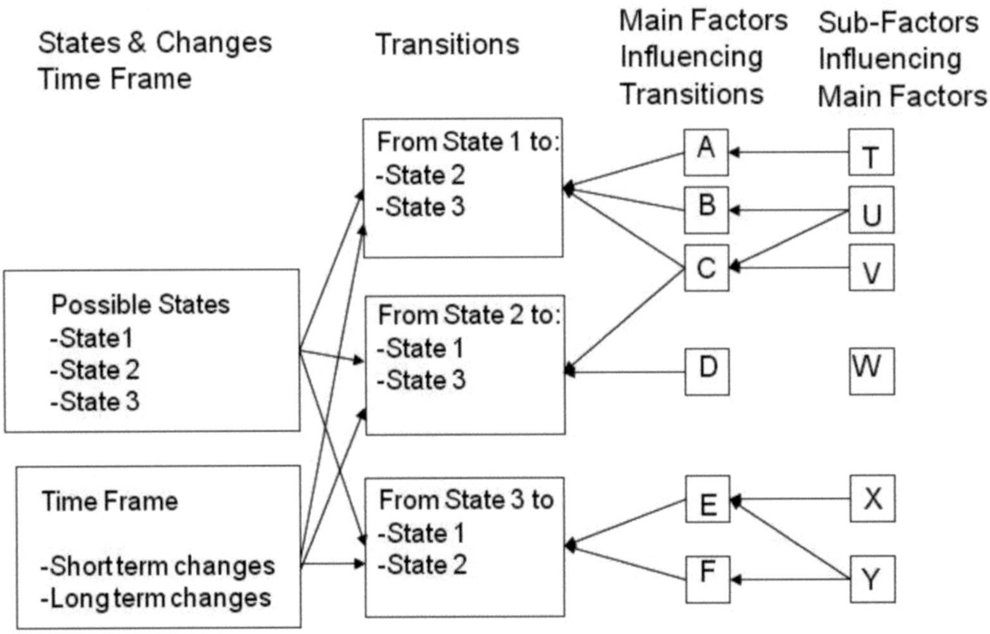

Fig. 3. This framework was used to construct Bayesian network structure from an STM (adapted from Bashari et al., 2009).

Next, states were defined for each node in the influence diagram. For the transition nodes, their states were the vegetation states in the STM. For the remaining nodes, that is the main factor and subfactor nodes, states were defined in consultation with the rangeland scientists who participated in building the STM. Figure 4 shows the completed influence diagram for the Steppe zone of Ghom and table 3 lists the states and the definitions for each node in the influence diagram.

Monitoring data or simulation models were not available to populate the influence diagram with conditional probabilities, so subjective probability estimates were obtained from the rangeland scientists who participated in building the STM.

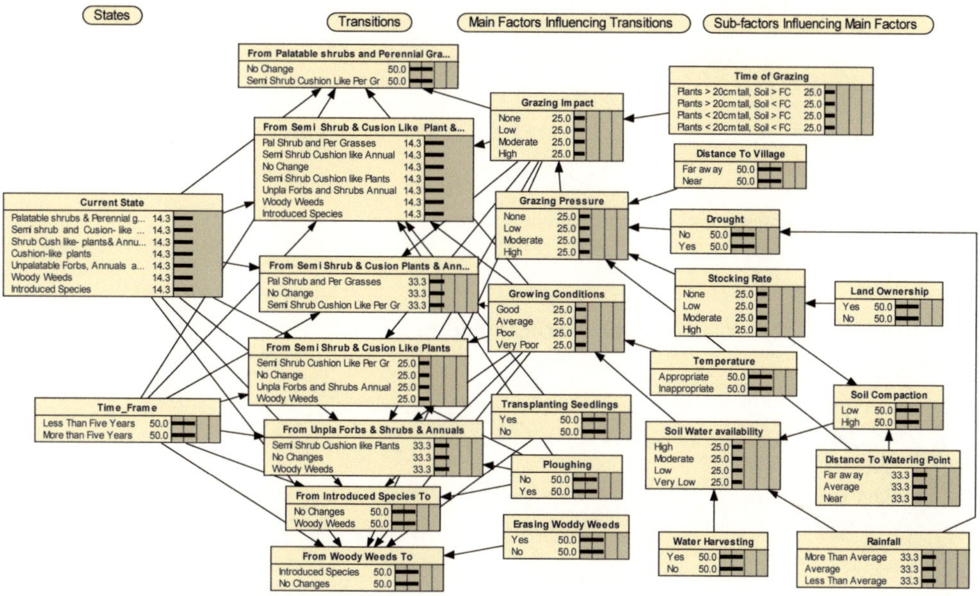

Fig. 4. Influence diagram of vegetation change for the Steppe zone of Ghom. Pal = Palatable, Per = Perennial, Unpla= Unpalatable, Gr = Grass, An = Annual

Node	States	Definition
	PSPG	Palatable shrubs & Perennial grasses (PSPG), Including high frequency of 3P grasses (Productive, Palatable, Perennial) and palatable shrubs such as *Salsola tomentosa* and *Andrachne sp*, it has highest production and stability;
Current State	SSCPPG	Semi-shrub and Cushion-like plants & Perennial grasses (SSCPPG). This state represents the overall condition in the Steppe zones and contains *Artemisia Sieberi* and *Stipa hohenackeriana* and some Cushion-like plants as dominant species;
	SCPA	Shrub, Cushion-like plants and Annuals (SCPA). Its composition is the same as SSCPPG but this state is highly preferred by sheep and goats due to of the abundant foliage of annuals. It is estimated this state can have up to 30% cover of annual grasses and forbs;
	SSCP	Semi-shrub & Cushion-like plants (SSCP). In this state, palatable shrubs such as *Salsola tomentosa* have disappeared and frequency of tall grass species (e.g. *Stipa hohenackeriana*) decreases dramatically. Erosion is high;
	UFAUS	Unpalatable forbs, Annuals and Unpalatable shrubs (UFAUS). This state represents the most degraded state. There are no perennial grasses and *Artmisia sieberi* has low frequency. The percentage cover is less than 10% and erosion is high;

Table 3. Continued

Node	States	Definition
	WW;	Woody Weeds (WW). Although it is a highly stable state; its species are not edible by livestock. In the early years after ploughing, the frequency of *Reseda sp* is higher but later on *Hulthemia persica*, a native weed of the Steppe zone, becomes dominant;
	IS	Introduced Species (IS). This state has two levels in which *Atriplex spp* constitutes the upper level, while various other species form the lower level of the vegetation structure. The percentage cover is low and the frequency of species such as *Artemisia sieberi* declines.
Time Frame	< five years > five years	Represent the likely years of transition under defined scenarios, Less than five years represents transitions occurring over short periods and more than five years represent transitions over longer periods of times (E.g. up to 10 or 20 years).
From PSPG to	No changes SSCPPG	The same as current state definition
From SSCPPG to	PSPG SCPA No change SSCP UFAUS WW IS	The same as current state definition
From SCPA to	PSPG No change SSCPPG	The same as current state definition
From SSCP	SSCPPG No change UFAUS WW	The same as current state definition
From UFAUS to	SSCP IS No change WW	The same as current state definition
From WW to	IS No change	The same as current state definition
From IS to	No change WW	The same as current state definition
Grazing Impact	None	None : when destocked
	Low	Low : when grazing pressure is low and grazing is during a time of range readiness
	Moderate	Moderate : when the plant is affected by grazing but the grazing pressure is in line with the carrying capacity and the appropriate time of grazing; if the grazing occurs when the range is not in a stage of readiness, even the low grazing pressure can have a moderate grazing impact
	High	High : when the grazing pressure is high and the grazing occurs when the range is not in a stage of readiness
Early Grazing	No Yes	Grazing rangeland prior to range readiness (e.g. grazing before grass plants reach the third leaf stage or grazing when soil is not dry enough to prevent damage to soil structure and plants)

Table 3. Continued

Node	States	Definition
Grazing pressure	None Low Moderate High	Represents the balance between how much the animals eat and how fast the pasture is growing. Grazing pressure (GP) = rate of removal of pasture / rate of supply of pasture. GP=0 None; GP<1, Low GP=1, Moderate, GP>1, High
Growing condition	Good	Good : when Soil Water Availability (SWA) is sufficient for plant growth and the temperature is appropriate
	Average	Average : when the SWA is average and temperature does not produce a major limitation for plant growth
	Poor	Poor : when the SWA is low or temperature causes some limitation for plant growth
	Very poor	Very Poor :when SWA is very low and/or temperature causes a major limitation for plant growth
Soil water availability	Hi	Amounts of soil moisture (SM) available to support plant growth; High = when the soil water content is above the wilting point for most of the growing season
	Average	Average: SM is available for 50 to 70% of the growing season
	Low	Low : SM is available for 20 to 50%of the growing season
	Very Low	Very Low : SM is available for less than 20% of the growing season
Transplan-ting Seedling	Yes No	Refers to whether seedlings of shrub species such as *Atriplex spp* are transplanted or not
Ploughing	No Yes	Refers to whether a site is ploughed or expansion of rainfed agriculture or not
Erasing Woody weeds	Yes No	Refers to use of appropriate mechanical or chemical treatment to control and eradicate woody weeds
Stocking rate	De-stocked	It describes how many animals a site can support. Destocked : using enclosures to keep the livestock out of a site
	Low	Low : the animal consumption is less than the available forage
	Moderate	Moderate : the animal consumption and available forage is equal
	High	High : the animal consumption is more than the available forage.
Distance to Village	Far away Near	Refers to the distance of the rangeland to the village, the closer the rangeland is to the village, the more likely it will be grazed by livestock
Drought	No Yes	Severe rainfall deficiencies over a year (there is a significant effect on vegetation when the rainfall is below 75% of the long term mean)
Soil Compac-tion	Low	Refers to the severity of soil compression. Low : good soil structure, only slight evidence of hard pans or surface crust.
	High	High : poor soil structure, evidence of hard pans and surface crust
Distance to watering point	Far away	Accessible area around watering points grazed heavily. Far away : >5 km away
	Average	Average : 1km to 5 km,
	Near	Near: < 1km
Water Harves-ting	Yes No	Determines whether water harvesting techniques such as contour furrow or pitting are used or not
Rainfall	High	High : > 150mm in areas at 1000 m above sea level and >200 mm in areas above 1300 m
	Average	Average : 150mm at 1000 m & 180mm at above 1300m
	Low	Low : <150 at 1000m & <180mm at above 1300 m
Tempera-ture	Appro-priate	Explains the temperature conditions that affect the phenological status of plants; Appropriate: no unseasonal temperatures occur;
	Inappro priate	Inappropriate : unseasonal temperatures occur and cause some damage to new growth.
Land ownership	Yes No	Explains the land tenure status: Yes:privately owned and used No : publicly used

Table 3. State definitions for nodes in the Steppe zone influence diagram (Fig. 4).

2.5 Making sense of the BBN model

Finally, the behaviour of the model was tested using scenario and sensitivity analysis. The results of the sensitivity analysis were returned to the Iranian rangeland scientists for review and feedback. If the scientists disagreed with the behaviour of the model, the conditional probability tables were revisited.

The sensitivity analysis revealed that grazing impact and growing condition were the two most important drivers of almost all transitions except for two (Table 4). "Grazing impact" represents the management influence on transitions and "growth condition" represents the environmental influence on transitions. These two had similar influences on most transitions. This result is supported by other studies in Iran, which suggest that frequent droughts coupled with mismanagement (e.g. overgrazing) combine to produce rapid land degradation (Nemati, 1986; Badripour, 2005). However, this result does not match the beliefs of governors or livestock managers. Most governors believe that grazing is the dominant factor responsible for rangeland degradation, while livestock managers believe that it is drought and growing conditions.

Drought and time of grazing had an effect on many transitions but only through their affect on grazing impact. High grazing impact allows the establishment of undesirable shrub species such as *Scariola orientalis* and *Noaea mucronata,* which compete heavily with favourable species for limited resources, especially water. Over-utilization with prolonged drought can reduce the tussock size of desirable perennial grasses, increasing the risk that they will be permanently lost from the rangeland seed bank. Unseasonal temperatures and low soil water availability increased the likelihood of poor or very poor growing conditions, making transitions to unpalatable forbs and annual states more likely.

The Steppe zone soil is generally low in organic matter. As a consequence, aggregate stability is low and the risk of soil compaction, surface sealing, and crust formation is high when overgrazing occurs, especially on silty soils. Hence, it is combination of poor soil characteristics and overgrazing that can lead to reduced rainfall effectiveness and soil water availability, triggering transitions to degraded states (Whisenant, 1999). In this case, water harvesting techniques are often needed to improve soil water availability and bring about transitions to palatable shrubs and perennial grasses.

The planting of seedlings was important in avoiding transition to introduced species states. The establishment of sown shrubs can also benefit from water harvesting techniques (Schreiber & Frasier, 1978). Nemati (1986) found water harvesting treatments for 5 years led to the recovery of *Artemisia sieberi, Stipa hohenackeriana, Aristida plumosa* , *Salsola spp.*, and *Astragalus siliquosus* in the Steppe zone. Irregular precipitation is the main reason for poor natural recruitment in rangelands and the establishment of sown rangelands in the Steppe zone (Monsen, 2002). It is therefore advisable to raise seedlings in a nursery and to transplant them prior to seasonal rains. Overgrazing, untimely grazing, drought and unseasonable temperatures can kill newly planted seedlings and thereby cause undesirable transitions.

Ploughing was an important driver of transitions to a state of woody weeds. Ploughing is a common cause for the establishment of woody weeds, such as *Reseda sp* and *Hulthemia persica,* in the Steppe zone. Ploughing often occurs near villages, not for cultivation or the expansion of rainfed agriculture, but to claim land ownership. Transitions away from woody weed are very expensive and require weed control plus the sowing of improved rangeland species such as *Atriplex spp* and *Eurotia ceratoides*. Spelling of rangeland is also required to allow sown rangelands to establish.

2.6 The modelling approach

BBN models have the ability to provide rangeland managers with decision support through their analytic capabilities. As mentioned before, two main types of analysis can be performed using a BBN, (a) prediction, and (b) diagnosis. Predictive analysis can be used to answer "what if" questions and diagnostic analysis can be used to answer "how" questions.

Figure 5 is an example of the Steppe zone of Ghom BBN used for predictions. Here, the selected states of input nodes (outer boxes) represent a scenario for a site. In Figure 5, the site is currently in the "Palatable shrubs and perennial grasses" state and the model is being used to predict the chance of a transition away from this state within a more than five years timeframe (note that the state "More than five years" is selected in the "Timeframe" node). The model shows that, under the selected scenario, the chance of transition away from "Palatable shrubs and perennial grasses" to "Semi shrub and cushion like plant" is relatively high (60.9%). The model also indicates the probable causes for this transition, that is, the probable high grazing impact (91.3%) and poor growing condition (62.4%). These causes were also highlighted by sensitivity analysis using the model (Table 4), which showed that the transition from "Palatable shrubs and perennial grasses" to "Semi shrub and cushion like plant" was most sensitive to grazing impact and growing condition. Table 5 shows the full conditional probability table "From palatable shrubs & perennial grasses" state.

Transition Number	Transition Name	Grazing impact	Growing condition	Ploughing	Transplanting seedlings	Erasing weeds
1	I, II	High	Moderate	*	*	*
2	II, I	Moderate	High	Low	High	*
3	II, III	Moderate	High	*	*	*
4	II, IV	High	Moderate	*	*	*
5	II, V	High	Moderate	*	*	*
6	II, VI	Moderate	Low	High	*	*
7	II, VII	Low	Moderate	*	High	*
8	III, I	Moderate	High	*	*	*
9	III, II	Moderate	High	*	*	*
10	IV, II	Moderate	High	*	*	*
11	IV, V	High	Moderate	*	*	*
12	IV, VI	Moderate	Low	High	*	*
13	V, IV	Moderate	High	*	*	*
14	V, VI	Moderate	Low	High	*	*
15	VI, VII	Very Low	Low	*	Moderate	High

(black)	(dark grey)	(light grey)	(white)	*
High influential	Moderate influential	Low influential	Very Low influential	None influential

An asterix (*) means that this factor had no influence on the transition.

Table 4. Summary of sensitivity analysis performed on the transition nodes in the Steppe BBN. The shading indicates the relative influence of factors on each transition, from most influential (black) to least influential (white).

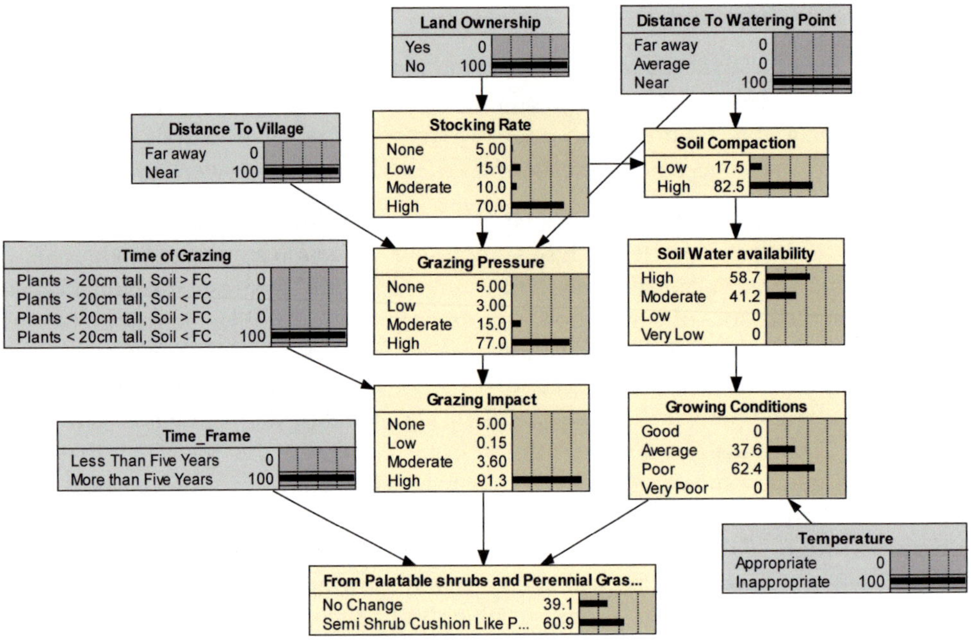

Fig. 5. Prediction using the Steppe zone of Ghom BBN.

Current State	Time_Frame	Grazing Impact	Growing Conditions	No Change	Semi Shrub Cushion Like Per Gr
Palatable shrubs & Perennial gras...	Less Than Five Years	None	Good	100.00	0.000
Palatable shrubs & Perennial gras...	Less Than Five Years	None	Average	98.000	2.000
Palatable shrubs & Perennial gras...	Less Than Five Years	None	Poor	95.000	5.000
Palatable shrubs & Perennial gras...	Less Than Five Years	None	Very Poor	90.000	10.000
Palatable shrubs & Perennial gras...	Less Than Five Years	Low	Good	100.00	0.000
Palatable shrubs & Perennial gras...	Less Than Five Years	Low	Average	95.000	5.000
Palatable shrubs & Perennial gras...	Less Than Five Years	Low	Poor	90.000	10.000
Palatable shrubs & Perennial gras...	Less Than Five Years	Low	Very Poor	80.000	20.000
Palatable shrubs & Perennial gras...	Less Than Five Years	Moderate	Good	90.000	10.000
Palatable shrubs & Perennial gras...	Less Than Five Years	Moderate	Average	80.000	20.000
Palatable shrubs & Perennial gras...	Less Than Five Years	Moderate	Poor	70.000	30.000
Palatable shrubs & Perennial gras...	Less Than Five Years	Moderate	Very Poor	60.000	40.000
Palatable shrubs & Perennial gras...	Less Than Five Years	High	Good	80.000	20.000
Palatable shrubs & Perennial gras...	Less Than Five Years	High	Average	70.000	30.000
Palatable shrubs & Perennial gras...	Less Than Five Years	High	Poor	40.000	60.000
Palatable shrubs & Perennial gras...	Less Than Five Years	High	Very Poor	30.000	70.000
Palatable shrubs & Perennial gras...	More than Five Years	None	Good	100.00	0.000
Palatable shrubs & Perennial gras...	More than Five Years	None	Average	98.000	2.000
Palatable shrubs & Perennial gras...	More than Five Years	None	Poor	95.000	5.000
Palatable shrubs & Perennial gras...	More than Five Years	None	Very Poor	90.000	10.000
Palatable shrubs & Perennial gras...	More than Five Years	Low	Good	100.00	0.000
Palatable shrubs & Perennial gras...	More than Five Years	Low	Average	95.000	5.000
Palatable shrubs & Perennial gras...	More than Five Years	Low	Poor	90.000	10.000
Palatable shrubs & Perennial gras...	More than Five Years	Low	Very Poor	80.000	20.000
Palatable shrubs & Perennial gras...	More than Five Years	Moderate	Good	90.000	10.000
Palatable shrubs & Perennial gras...	More than Five Years	Moderate	Average	80.000	20.000
Palatable shrubs & Perennial gras...	More than Five Years	Moderate	Poor	60.000	40.000
Palatable shrubs & Perennial gras...	More than Five Years	Moderate	Very Poor	50.000	50.000
Palatable shrubs & Perennial gras...	More than Five Years	High	Good	70.000	30.000
Palatable shrubs & Perennial gras...	More than Five Years	High	Average	60.000	40.000
Palatable shrubs & Perennial gras...	More than Five Years	High	Poor	20.000	80.000
Palatable shrubs & Perennial gras...	More than Five Years	High	Very Poor	0.000	100.00

Table 5. Full conditional probability table for "From palatable shrubs & perennial grasses" state relating "Time frame", "Grazing impact" and "Growing conditions" to possible transitions. In this example, probabilities for the first row is read from the table as, when current state is "Palatable shrubs and perennial grasses", "Time frame" is less than five years, "Grazing impact" is none and "Growing condition" is good, there is 100 % chance of "No changes" and 0% chance of a transition to "Semi shrub cushion-like plant & perennial grasses".

Besides answering the "what if" questions the BBN model can also help to answer "how" questions. For example, how might a manager move from an "Semi shrub and cushion like plants" to a "Palatable shrubs and perennial grasses"? Figure 6 is an example of the Steppe zone of Ghom BBN being used to answer this question using diagnosis. The model shows that within a less than five year time frame, this transition is most likely if there is no grazing impact and also good growing condition (see the "Grazing impact" and "Growing condition" nodes), and this is most likely to be achieved by destocking (see the "Stocking rate" node). The model also shows that, more than average rainfall and appropriate temperature are important to achieving good growing condition (see the more than average in the rainfall and appropriate for temperature nodes).

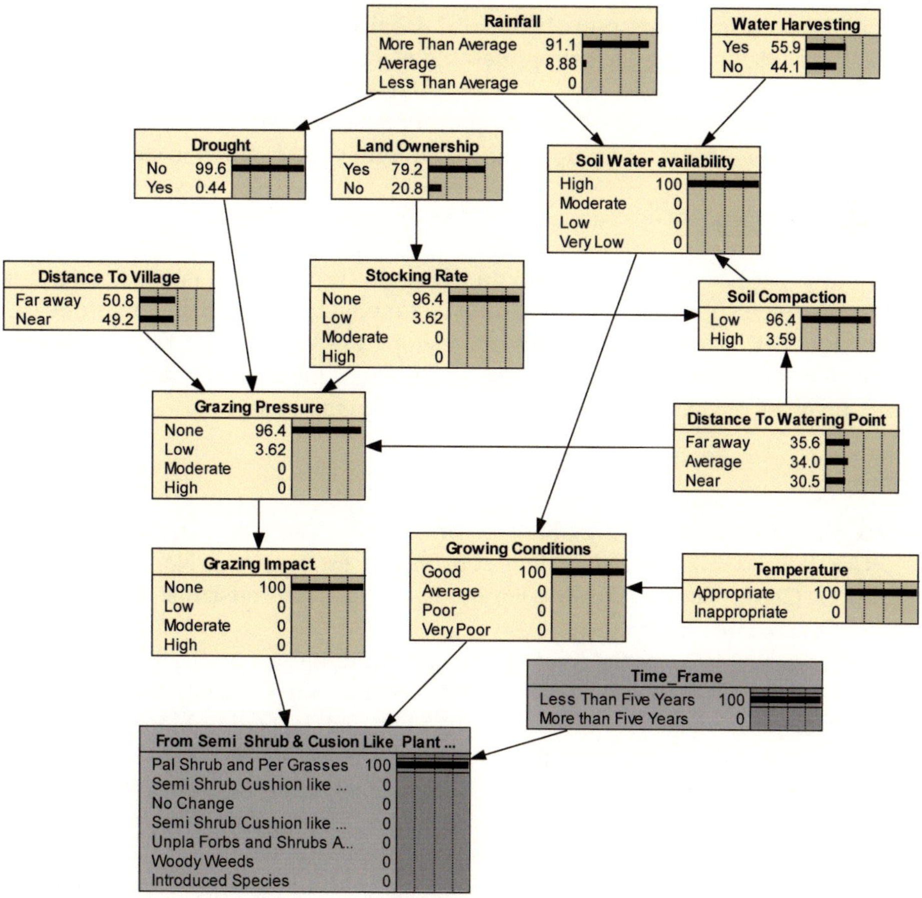

Fig. 6. Using the model for diagnostic assessment to determine the most likely causes of a transition.

3. Conclusion

The methodology used in this chapter (integrating the STM with the BBN) can provide a useful approach to accommodate uncertainty in highly uncertain systems (e.g. Iranian rangeland). Despite the advantage of STMs, they are traditionally descriptive diagrams and are unable to be used for predictive modelling and scenario analysis. They also handle uncertainty associated with causes of vegetation change poorly. Bayesian Belief Network (BBN) used in this study assist in the development of a dynamic and predictive STM by providing a graphical modelling framework for building a probability-based cause and effect model. The results indicated that the BBN approach is a highly useful mechanism for adding value to descriptive STMs. First, it allowed the uncertainty in transitions to be expressed by using probabilistic relationships. Second, the approach provided a scenario and sensitivity analysis tool for both scientists and landholders to assess the probable vegetation outcomes of rangeland management decisions, and to identify those management options most likely to improve or degrade vegetation condition. Third, it is particularly complementary to the adaptive management process, because monitoring records can be used to update probability relationships within the BBN model over time. Therefore, the modelling approach supported the planning, monitoring and review steps of the adaptive management cycle. This is an advantage over current rangeland management simulation models that are good at supporting management planning through their predictive capabilities, but poor at supporting monitoring and evaluation steps.

4. Acknowledgment

I am grateful to Mr Seied Mehdi Adnani and Hossein Bagheri from natural resources and animal affairs research centre of Ghom province and Dr. Hossein Barani from Gorgan University for their participation in the workshops associated with this study and the sharing of their expert knowledge and also to professor Ockie Bosch and Dr. Carl Smith from the University of Queensland for their assistance in this study.

5. References

Allen-Diaz, B. & Bartolome, J.W. (1998). Sage Brush-Grass Vegetation Dynamics: Comparing Classical and State Transition Model. *Ecological Application*, vol.8, No. 3, pp. 795-804.

Badripour, H. (2005). Country Pasture,Forage Resource Profiles-Islamic Republic of Iran, FAO, Available From http://www.fao.org/ag/AGP/AGPC/doc/Counprof/Iran/Iran.htm

Bashari, H.; Smith, C. & Bosch, O.j.H. (2009). Developing Decision Support Tools for Rangeland management by Combining State and Transition Models and Bayesian Belief Networks, *Agricultural Systems*, Vol. 99, No.1, pp: 23- 34.

Bosch, O.J.H., Ross, A.H. & Beeton, R.J.S. (2003). Integrating Science and Management through Collaborative Learning and Better Information Management. *Systems Research and Behavioral Science System*, Vol. 20, pp. 107-18.

Briske, D.D, Fublendorf, S.D & Smeins, F.E (2005).State and Transition Models, Thresholds, and Rangeland Health: a Synthesis of Ecological Concepts and Prespectives. *Rangeland Ecology & Management*, Vol. 58, No. 1, pp. 1-10.

Carlson, D.H., Thurow, T.L. & Jones, C.A. (1993).Biophysical Simulation Support Models as a Foundation of Decision Support Systems, In: *Decision Support Systems for the management of Grazing Lands*, J.W. Stuth, & B.G. Lyons (Ed.), , UNESCO and The Parthenon Publishing Group, Carnforth, UK.

Friedel,M. (1991).Range Condition Assessment and the Concept of Thresholds: a Viewpoint. *Journal of Range management*, vol. 44, No. 5, pp. 422-6.

Gross, J., & McAllister, R. (2003). Modelling Rangeland as Complex Adaptive Social-Ecological Systems, *proceedings of VIIth International Rangeland Congress*, Durban,South Africa.

Gross, J.E., McAllister, R.R.J., Abel, N., Smith, D.M.S., & Maru, Y. (2006).Australian Rangelands as Ccomplex Adaptive Systems: A Conceptual Model and Preliminary Results. *Environmental Modelling & Software*, vol. 21, No. 9, pp. 1264-72.

Hall, T.J., Filet, P.G., Banks, B., & Silcock, R.G. (1994). A State and Transition Model of the Aristida-Bothriochloa Pasture Community of Central and Southern Queensland. *Tropical Grassland*, Vol. 28, pp. 270-3.

Jansen, F.V. (1996). *An Introduction to Bayesian Networks*, Springer Verlag, New York.

Knapp, C.N.; Fernandez-Gimenez, M.; Kachergis, E. & Rudeen, A. (2011). Using Participatory Workshops to Integrate State-and-Transition Models Created With Local Knowledge and Ecological Data. *Rangeland Ecology & Management*, Vol. 64, No. 2, pp. 158-170.

Laycock, W.A. (1991). Stable States and Thresholds of Range Condition on North American Rangeland: A View Point. *Journal of Range management*, vol. 44, No. 5, pp. 427-33.

Loehle, C. (2004). Challenges of Ecological Complexity. *Ecological complexity*, Vol. 1, No. 1, pp. 3-6.

Monsen, S.B. (2002).Ecotypic Variability, Seed Features, and Seedbed Requirements of Big Sagebrush., *Paper Presented to Restoration and Management of Sagebrush/ Grass Communities Workshop*, Elko, Nevada.

Morghan, K.J.R., Sheley, R.L., & Svejcar, T.J. (2006).Successful Adaptive Management-the Integration of Research and Management. *Rangeland ecology and management*, Vol. 59, No. 2, pp. 216-9.

National Land & Water Resources Audit (2004). Natural Resources Models in the Rangelands, *a Review Undertaken for the National Land and Water Resources Audit*, CSIRO Sustainable Ecosystems, Brisbane.

Nemati, N. (1977).Range rehabilitation Problems of the Steppic Zone of Iran. *Journal of Range Management*, Vol. 30, No. 5, pp. 339-42.

Nemati, N. (1986). Pasture Improvement and Management in Arid Zones of Iran. *Journal of Arid Environments*, Vol. 11, No. 1, pp. 27-35.

Phelps, D.G., & Bosch, O.J.H. (2002). A Quantitative State and Transition Model for the Mitchell Grasslands of Central Western Queensland. *Rangeland*, Vol. 24, No. 2, pp. 242-67.

Pielke Jr, R.A. (2001). Room for Doubt. *Nature*, Vol. 410, pp. 151.

Pielke Jr, R.A.(2003).The Role of Models in Prediction for Decision, In: *Model in ecosystem science*, C.D. Canham; J.J. Cole, & W.K. Lauenroth (Ed.), Princeton University Press, New Jersey.

Provenza, F.D. (1991).View Point: Range Science and Range Management are Complementary but Distinct Endeavours. *Journal of Range Management*, Vol. 44, No. 2, pp. 181-3.

Ringold, P.L.; Alegaria, J.; Czaplewski, R.L; Mulder, B.S.; Tolie, T. & Burnett, K. (1996). Adaptive monitoring Design for Ecosystem Management, *Ecological Application*, Vol. 6, No. 3, pp. 745-7.

Sabine, E.; Schreiber, G.; Bearlin, A.R.;Nicol, S.J. & Todd, C.R. (2004). Adaptive Management: a Synthesis of Current Understanding and Effective Application. *Ecological Management & Restoration*, Vol. 5, No. 3, pp. 177-82.

Schreiber, H.A. & Frasier, G.W. (1978). Increasing rangeland Forage Production by Water Harvesting, *Range Management*, Vol. 31, No. 1, pp. 37-40.

SRM, (1989). *A Glossary of Terms used in Rrange Management. Third Ed.*, Soc. Range Manage, Denver, Colorado, U.S.A.

Tueller, P. (1988). *Vegetation Science Applications for Rangeland Analysis and Management*, Kluwer Academic.

Vayssieres, M.P. & Plant, R.E. (1998). *Identification of Vegetation State- and- Transition Domains in California's Hardwood Rangelands.*, California Department of Forestry and Fire Protection, California.

Walker, B.H. & Janseen, M.A. (2002). Rangelands, Pastoralists and Governments: Interlinked Systems of People and Nature. *Royal society*, Vol. 357, pp. 719-25.

Westoby, M.; Walker, B. & Noy-Meir, I. (1989). Opportunistic Management for Rangelands not at Equilibrium', *Journal of Range Management*, Vol. 42, No. 4, pp. 266-74.

Whisenant, S.G. (1999). *Repairing Damaged Wildlands*, Cambridge Universtity Press, Cambridge.

2

Building a Bayesian Network Model Based on the Combination of Structure Learning Algorithms and Weighting Expert Opinions Scheme

Wichian Premchaiswadi and Nipat Jongsawat
Graduate School of Information Technology in Business, Siam University
Thailand

1. Introduction

Bayesian networks (BNs) is probabilistic graphical models that are widely used for building expert systems in several application domains. In the context of expert systems, either probabilistic or heuristic, the development of explanation facilities is important for three main reasons. First, the construction of those systems with the help of human experts is a difficult and time consuming task, and prone to errors and omissions. A Bayesian network tool can help the knowledge engineers and experts who are taking part in the project to debug the system when it does not yield the expected results and even before a malfunction occurs. Second, human beings are reluctant to accept the advice that is offered by a machine if they are not able to understand how the system arrived at those recommendations. Third, an expert system that is used as an intelligent tutor must be able to communicate to the apprentice the knowledge it contains, the way in which the knowledge has been applied to arrive at a conclusion, and what would have happened if the user had introduced different pieces of evidence (what-if reasoning). One of the most difficult obstacles in the practical application of probabilistic methods is the effort that is required for model building and, in particular, for quantifying graphical models with numerical probabilities. The construction of Bayesian Networks (BNs) with the help of human experts is a difficult and time consuming task, which is prone to errors and omissions especially when the problems are very complicated or there are numerous variables involved. Learning the structure of a BN model and causal relations from a dataset or database is important for extensive BNs analysis. In general, the causal structure and the numerical parameters of a BN can be obtained using two distinct approaches. First, they can be obtained from an expert. Second, they can also be learned from a data set. The main drawback of the first approach is that sometimes there is not enough causal knowledge to establish the structure of the network model with certainty and estimation of probabilities required for a typical application is a time-consuming task because of the number of parameters required (typically hundreds or even thousands of values). Thus, the second approach can initially help human experts or a group of experts build a BN model and they can make it applicable at a later time. In practice, some combination of these two approaches is typically used.

This article presents a SMILEBN web application for building a Bayesian network model. The SMILEBN can build a BN model based on using two techniques: 1) to build a BN model by applying the structure learning algorithms to a dataset, and 2) to use group decision making technique for weighting the degree of an expert's opinion in identifying influential effects from parent variables to child variables in the model. Finally, the BN model which all the experts agree to use is obtained. In case that the BN model which is built from a data set is complex, the SMILEBN users can set a threshold value for the model in order to minimize the number of relationships among the nodes in the BN model. When the number of relationships among the nodes decreases, the complexity of the conditional probability table on each child node also decreases.

This article is organized as follows: Section 2 addresses related work. Section 3 presents the tools that are used to build a BN causal structure from a dataset. Section 4 presents the method to use group decision making technique for weighting the degree of an expert's opinion in identifying influential effects from parent variables to child variables. Section 5 presents a SMILEBN web application. Section 6 presents a conclusion and discusses some perspectives and ideas for future work.

2. Related work

There are various kinds of software applications that can be used to create decision theoretic models, learn the causal structure, and perform diagnosis based on BNs. There are both commercial and non-commercial software applications available. The commercial software applications are widely used in a business environment. Many of them are integrated into business analysis software and used particularly for solving difficult business problems. The non-commercial software applications are extensively used for the educational purposes. This article reviews only the most relevant subset of non-commercial software applications based on BNs.

B-Course is an analysis tool that was developed in the fields of Bayesian and causal modelling (Mylltmaki et al., 2002). It is a free web-based online data analysis tool, which allows users to analyze data for multivariate probabilistic dependencies. It also offers facilities for inferring certain type of causal dependencies from the data. B-Course is used via a web-browser, and requires the user's data to be a text file with data presented in a tabular format typical for any statistical package (e.g., SPSS, Excel text format). It offers a simple three step procedure (data upload, model search, and analysis of the model) for building a BN dependency model. After searching the model, B-Course provides the best model to the user via a report. Users can continue to search for the next best model but they must make the decision for selecting the best model that fits their needs. Selecting the best model is sometimes very difficult for inexperienced users. In B-Course, there are no structural learning algorithms provided for the user to aid in selection. The analysis method, modelling assumptions, restrictions, model search algorithms, and parameter settings are totally transparent to the user.

Elvira is a tool for building and evaluating graphical probabilistic models (Lacave et al., 2007). It is a non web-based application. It is implemented in Java, so that it can run on different platforms. It contains a graphical interface for editing networks, with specific options for canonical models (e.g., OR, AND, MAX, etc.), exact and approximate algorithms for discrete and continuous variables, explanation facilities, learning methods for building networks from databases, algorithms for fusing networks, etc. Elvira is structured as four main modules: (1) data representation- containing the definition of the data structures that are needed for managing BNs and IDs in Java, (2) data acquisition- including the classes that are necessary for saving and loading a network from either a file or a database, (3) processing - implementing the algorithms for processing and evaluating models, and (4) visualization - defining the Elvira graphical user interface (GUI) which obviously makes use of the classes that are included in the previous modules.

GeNIe (Graphical Network Interface) is a versatile and user friendly development environment for building graphical decision models (Druzdzel, 1999). The original interface was designed for a Structural Modeling, Inference, and Learning Engine (SMILE). GeNIe may be seen as an outer shell to SMILE. GeNIe is implemented in Visual C++ and draws heavily on the Microsoft foundation classes. GeNIe provides numerous tools for users such as an interface to build Bayesian network models or influence diagrams, to learn the causal relationships of a model using various algorithms, and to perform model diagnosis. In order to use GeNIe efficiently, the GeNIe software must be installed and the user should have some background knowledge about probabilistic graphical models and become familiar with the tools provided in GeNIe.

Poompuang, et al presents a development environment for building graphical decision-theoretic models based on BNs and influence diagrams working on the website by utilizing an original engine called "SMILE" (Poompuang, et al., 2007). They propose the idea of building and developing graphical decision-theoretic models on a web page in order to overcome such the limitation of Bayesian belief network software developed on a windows-based platform, which makes the models not easily portable and is limited in its graphical representation across multiple system platforms. They present a prototype of BN models and influence diagrams in a World Wide Web environment, which can be displayed by a standard web browser.

Tungkasthan, et al presents a visualization of BN and influence Diagram models on a website (Tungkasthan et al., 2008). They develop an application based on the Macromedia Flash and Flash Remoting technologies. The application model on the client side is constructed by using the Macromedia Flash and the connection between a client and web server is developed by using the Flash Remoting technology. They use the capability of Marcomedia Flash and Flash Remoting technology to build richer, more interactive, more efficient, and more intuitive user interfaces for their applications than are possible with other web technologies such as JSP and Java applets. Their applications also provide a powerful, intuitive drag-and-drop graphical authoring tool that is comfortable for the users and have quick-loading and dynamic interfaces.

Jongsawat, et al presents a SMILE web-based interface that permits users to build a BN causal structure from a dataset or database and perform Bayesian network diagnosis through the web (Jongsawat & Premchaiswadi, 2009). There are several learning algorithms such as Greedy Thick Thinning, PC, Essential Graph Search, and Naive Bayes provided for the user. The user can just select the desired learning algorithm and adjust its parameter settings to learn the model structure. After building the BN structure, the user is able to quantify uncertain interactions among random variables by setting observations (evidence) and use this quantification to determine the impact of the observations. The SMILE web-based interface was developed based on SMILE, SMILEarn, and SMILE.NET. It uses a novel, user-friendly interface which interweaves the steps in the BN analysis with brief support instructions on the web page. They also present a technique to dynamically feed data into a diagnostic BN model and a web-based user interface for the models (Jongsawat et al., 2010). In their work, the BN model (the students' attitude towards several factors in a college enrolment decision) is fixed and the data obtained from an online questionnaire are saved into a database and transferred to the model. The user can observe the changes in the probability values and the impact the changes have on each node in real-time after clicking on a belief update button. Users can also perform Bayesian inference in the model and they can compute the impact by observing values of a subset of the model variables on the probability distribution over the remaining variables based on real-time data. They also present a methodology based on group decision making for weighting expert opinions or the degree of an expert's belief in identifying the causal relationships between variables in a BN model (Jongsawat et al., 2010).

3. Tools to build a bayesian network causal structure from a dataset

The core reasoning engines of the web-based interface development capability proposed in this article consist of SMILE (Structural Modeling, Inference, and Learning Engine), SMILEarn, and JSMILE. SMILE is a reasoning engine that is used for graphical probabilistic models and provides functionality to perform diagnosis. SMILEarn is used for obtaining data from a data source, pre-processing the data, and learning the causal structure of BN models. JSMILE is used for accessing the SMILE library from the web-based interface. This section provides some more detailed information about SMILE, SMILEarn and JSMILE wrapper.

SMILE is a fully platform independent library of functions implementing graphical probabilistic and decision-theoretic models, such as Bayesian networks, influence diagrams (IDs), and structural equation models (Druzdzel, 1999). Its individual functions, defined in the SMILE Application Programmer Interface (API), allow creating, editing, saving, and loading graphical models, and using them for probabilistic reasoning and decision making under uncertainty. SMILE can be embedded in programs that use graphical probabilistic models as their reasoning engines. Models developed in SMILE can be equipped with a user interface that best suits the user of the resulting application. SMILE is written in C++ in a platform-independent manner and is fully portable. Model building and the reasoning process are under full control of the application program as the SMILE library serves merely as a set of tools and structures that facilitates them.

SMILEarn extends the functionality provided by SMILE. It provides a set of specialized classes that implement learning algorithms and other useful tools for automatically building graphical models from data. It is a C++ library that contains a set of data structures, classes, and functions that implement learning algorithms for graphical models and includes other functionality (such as data access, storage and pre-processing) that can be used in a model in conjunction with SMILE. Although SMILEarn is a module of SMILE, which means that it requires SMILE to be used, but one can use SMILE without the need to install and use SMILEarn.

JSMILE is a library of java classes for reasoning about graphical probabilistic models, such as Bayesian networks and influence diagrams. It can be embedded in programs that use graphical probabilistic models as a reasoning engine. It is a wrapper library that enables access to the SMILE and SMILEXML C++ libraries from java applications. JSMILE is not limited to stand-alone applications. It can also be used on the back-end side of a multi-tiered application.

4. Weighting expert opinions scheme

We apply the weighting expert opinions scheme to the BN model, which is constructed based on the core reasoning engines mentioned in previous section. In this section we present the sequence of steps in the decision making procedure using the weighting expert opinions scheme. The sequence of decision procedure is described as follows.

Let $V = \{v_1,\ldots,v_m\}$ be a set of decision makers (or experts) who present their opinions on the pairs of a set of alternatives $X = \{x_1,\ldots,x_n\}$ where m is the number of experts and n is the number of alternatives in a set. Both m and n must be greater than or equal to 3; $m, n \geq 3$. P(V) denotes the power set of $V(I \in P(V))$. Linear orders are binary relations satisfying reflexivity, antisymmetry and transitivity, and weak orders (or complete preorders) are complete and transitive binary relations. With $|I|$ we denote the cardinality of I.

We consider that each expert classifies the alternatives within a set of linguistic categories $L = \{l_1,\ldots,l_q\}$, with $q \geq 2$, linearly ordered $l_1 > l_2 > \ldots > l_q$ (Herrera, 2000: Yager, 1993). The individual assignment of each expert v_i is a mapping $C_i = X \rightarrow L$ which assigns a linguistic category $C_i(x_u) \in L$ to each alternative $x_u \in X$. Associated with C_i, we consider the weak order R_i defined by $x_u R_i x_v$ if $C_i(x_u) \geq C_i(x_v)$. It is important to note that experts are not totally free in declaring preferences. They have to adjust their opinions to the set of linguistics categories, so the associated weak orders depend on the way they sort the alternatives within the fixed scheme provided by $L = \{l_1,\ldots,l_q\}$. For instance, for q = 5 expert-1 can associate the assignment: $C_1(x_3) = l_1$, $C_1(x_1) = C_1(x_2) = C_1(x_4) = l_2$, $C_1(x_5) = l_3$, $C_1(x_6) = C_1(x_7) = l_4$, $C_1(x_8) = C_1(x_9) = l_5$; expert 2 can associate the assignment: $C_2(x_1) = l_1$, $C_2(x_4) = l_2$, $C_2(x_5) = l_3$, $C_2(x_7) = C_2(x_8) = l_4$, $C_2(x_2) = C_2(x_3) = C_2(x_6) = l_5$; and so on. A profile is a vector $C = (C_1,\ldots,C_m)$ of individual assignments. We denote by C the set of profile.

Every linguistic category $l_k \in L$ has associated a score $s_k \in R$ in such a way that $s_1 \geq s_2 \geq \ldots \geq s_p$. For the expert v_i, let $S_i \rightarrow R$ be the mapping which assigns the score to each alternative, $S_i(x_u) = s_k$ whenever $C_i(x_u) = l_k$. The scoring vector of v_i is $(S_i(x_1),\ldots,S_i(x_n))$.

Naturally, if $s_i > s_j$ for all $i, j \in \{1,\ldots,q\}$ such that $i > j$, then each linguistic category is determined by its associated score. Thus, given the scoring vector of an expert we directly know the way this individual sorted the alternatives. Although linguistic categories are equivalent to decreasing sequences of scores, there exist clear differences from a behavioral point of view.

4.1 Sort the alternatives and assign a score

Experts $\{v_1,\ldots,v_m\}$ sort the alternatives of $X = \{x_1,\ldots,x_n\}$ according to the linguistic categories of $L = \{l_1,\ldots,l_q\}$. Then, we obtain individual weak orders $R_1,\ldots,R_m$ which ranks the alternatives within the fixed set of linguistic categories. Next, taking into account the scores

$s_1,\ldots,s_p$ associated with $l_1,\ldots,l_q$, a score is assigned to each alternative for every expert: $S_i(x_u)$, $I = 1,\ldots m$; $u = 1,\ldots,n$.

4.2 Calculate the euclidean distance

In order to have some information about the agreement in each subset of experts, we first calculate a distance between pairs of preferences (scoring vector). Since the arithmetic mean minimizes the sum of distances to individual values with respect to the Euclidean metric, it seems reasonable to use this metric for measuring the distance among scoring vectors. Let $(S(x_1),\ldots,S(x_n))$ and $(S'(x_1),\ldots,S'(x_n))$ be two individual or collective scoring vectors. The distance between these vectors by means of the Euclidean metric is derived by (1).

$$d(s,s') = \sqrt{\sum_{u=1}^{n} (s(x_u) - S'(x_u))^2} \quad (1)$$

4.3 Aggregate the expert opinions

We aggregate the expert opinions by means of collective scores which are defined as the average of the individual scores. There are several steps in this procedure.

4.3.1 Calculate the overall agreement measure

We calculate a specific agreement measure which is based on the distances among individual and collective scoring vectors in each subset of experts. The overall agreement measure is derived by (2).

$$M(C_r I) = 1 - \frac{\sum_{V_i \in I} d(S_i, S)}{|I| S_1 \sqrt{n}} \quad (2)$$

We note that $S_1\sqrt{n}$ is the maximum distance among scoring vectors, clearly between $(S(x_1),\ldots, S(x_n)) = (s_1,\ldots, s_1)$ and $(S'(x_1),\ldots, S'(x_n)) = (0,\ldots,0)$; $d(S, S') = S_1\sqrt{n}$. $M(C, I)$ is equal to 0 if $I = \phi$. Then, $M(C, I) \in [0, 1]$, for every $(C, I) \in C \times P(V)$. It is easy to see that the overall agreement measure satisfies the other axioms of (Bosch, 2005), Anonymity and Neutrality.

4.3.2 Calculate the overall contribution to the agreement

We now calculate an index which measures the overall contribution to agreement by each expert with respect to a fixed profile, by adding up the marginal contributions to the agreement in all subsets of experts. The overall contribution to the agreement of expert v_i with respect to a profile is defined by (3).

$$W_i = \sum_{I \subset V} M(C,I) - M(C, I \setminus |v_i|)) \quad (3)$$

If $w_i > 0$, we can conclude that expert v_i positively contributes to the agreement; and if $w_i < 0$, we can conclude that that expert v_i negatively contributes to the agreement.

4.3.3 Calculate the weak order

We now introduce a new collective preference by weighting the score which experts indirectly assign to alternatives with the corresponding overall contribution to the agreement indices. The collective weak order associated with the weighting vector w = $(w_1,\dots,w_m)$, R^w, is defined by (4) and (5).

$$W_i = \sum_{I \subset V} (M(C,I) - M(C,I \setminus |v_i|)) \tag{4}$$

where

$$S^w(x_u) = \frac{1}{m}\sum_{i=1}^{m} w_i S_i(x_u) \tag{5}$$

Consequently, we prioritize the experts in order of their contribution to agreement (Cook et al., 1996).

5. SMILEBN web application

The following steps in this section describe how a SMILEBN web application works for creating the BN models based on the combination of structure learning algorithms and weighting expert opinions scheme. The structure of the proposed framework is presented in Fig. 1. It shows a practical framework for building diagnostic Bayesian networks based on both learning algorithms and expert beliefs.

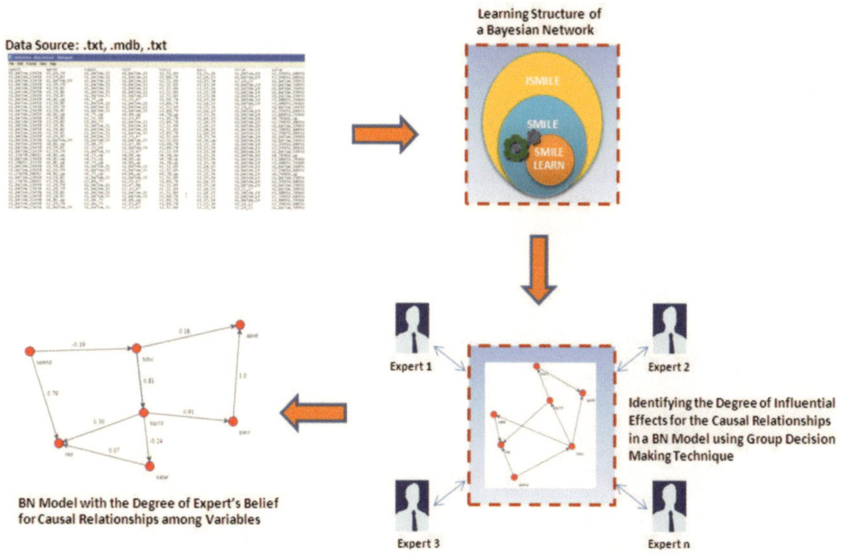

Fig. 1. A Practical framework for building diagnostic Bayesian networks based on both learning algorithms and expert beliefs

The first step is to import the data from a database or the data stored in the text file to the SMILEBN web application. Users select the file from the list and then clicks on "OK" button.

SMILEBN uses the data grid view to display the loaded data files and let's users work with them much like with spreadsheets. If the data file does not contain any missing values, SMILEBN will inform the users about that and "Next" button will be enabled. Otherwise, SMILEBN will tell how many rows were selected and the corresponding ones will become highlighted in the data grid. Users must solve the missing values manually (See Fig. 2 and Fig. 3). Once they have a data set prepared they can proceed to learning the network by picking the method and setting it's parameters. Note that if the data set contains continuous variables they will need to be discretized for some learning methods to be able to run, e.g. Naive (See Fig. 4 and Fig. 5). Fig. 6 shows the structure of a Bayesian network after applying the learning process. It shows the probability values over all nodes after performing Bayesian updating or belief updating (by clicking on "Update Belief" button) when the users move the mouse cursor over any node (See Fig.7). The user is allowed to perform a model diagnosis by entering observations (evidence) for some of the context and evidence variables. Fig. 8 shows the screenshot of the BN model diagnosis. The user begins the BN model diagnosis by performing a right click on a node and selects the state for setting the evidence for the test.

After setting the evidence, they click on the "Update Belief" button to update the model. Fig. 2 - Fig. 8 mainly show the methods to build a BN model in the SMILEBN web application based on the structure learning algorithms mentioned in section 3. Next the weighting expert opinions scheme will be applied to the BN model.

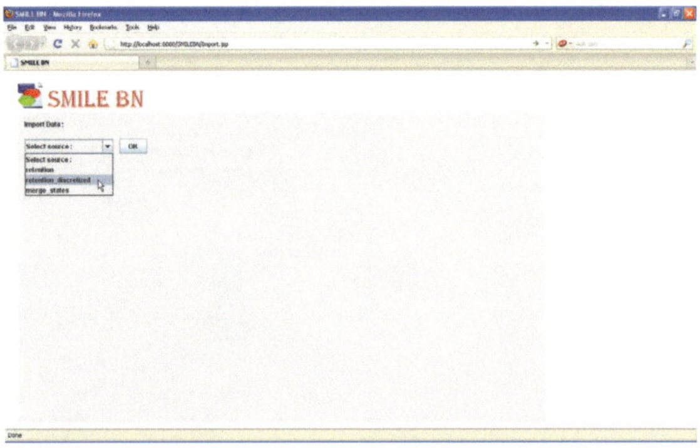

Fig. 2. Importing the data files

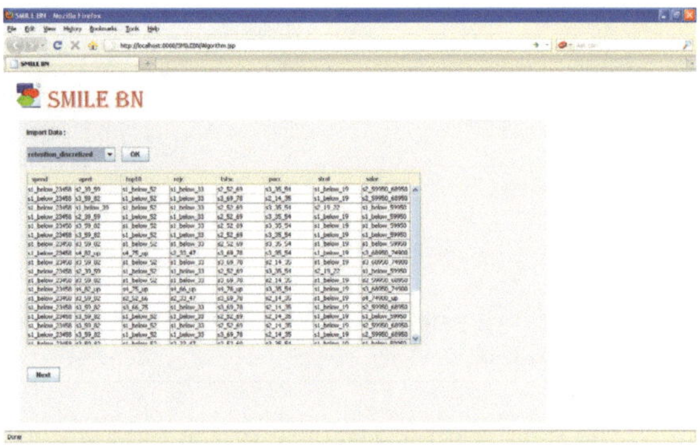

Fig. 3. Data grid view to display the loaded data files

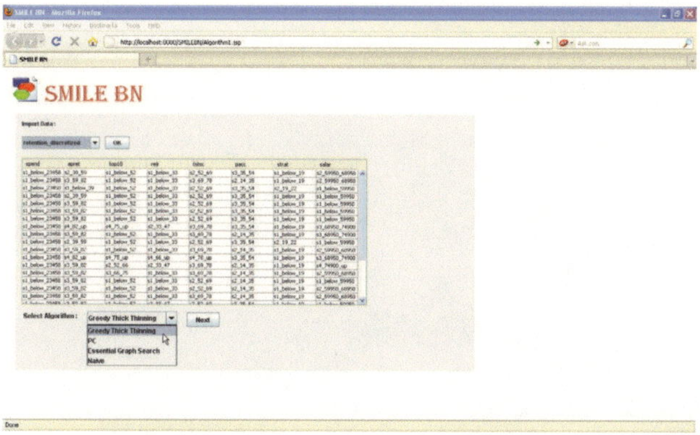

Fig. 4. Selecting the learning algorithms

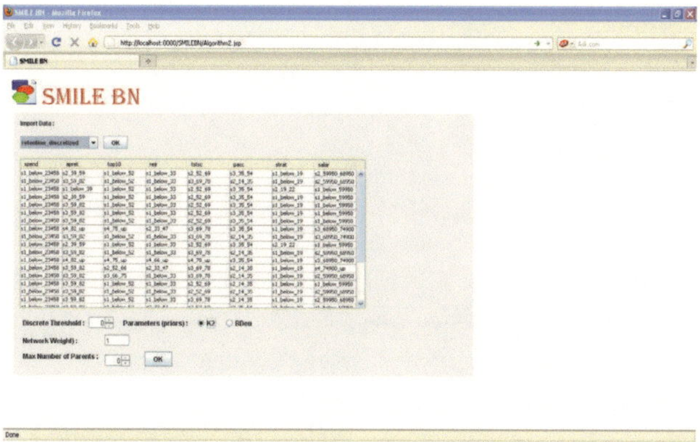

Fig. 5. Setting learning algorithm's parameter

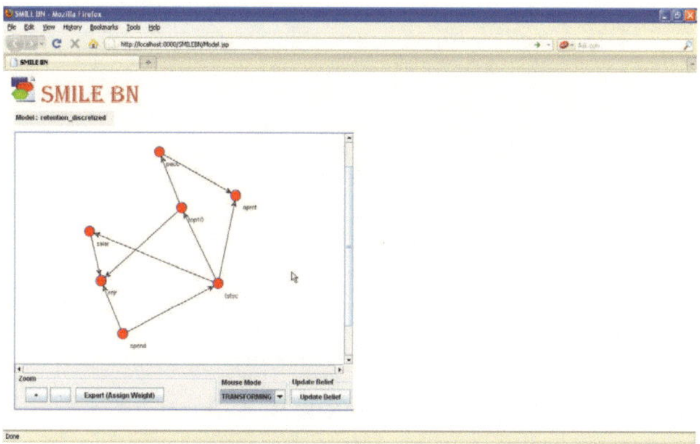

Fig. 6. The structure of a Bayesian network after applying the learning process

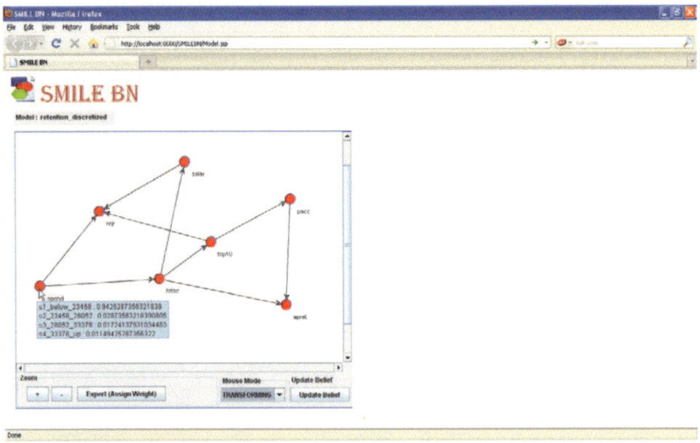

Fig. 7. Bayesian network with the probability values over all nodes after performing Bayesian updating (belief updating)

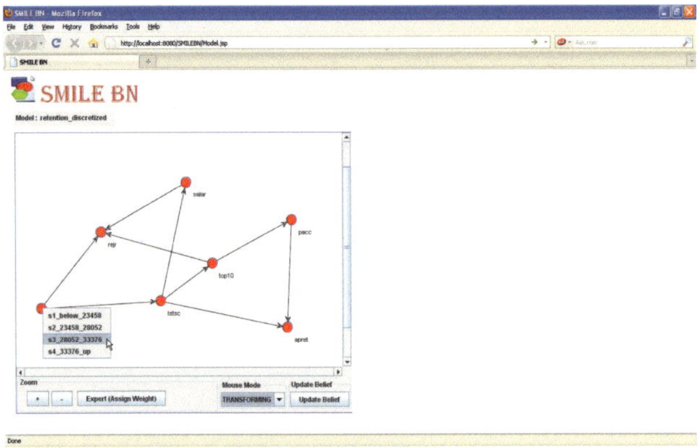

Fig. 8. Selecting an evidence corresponding to the node

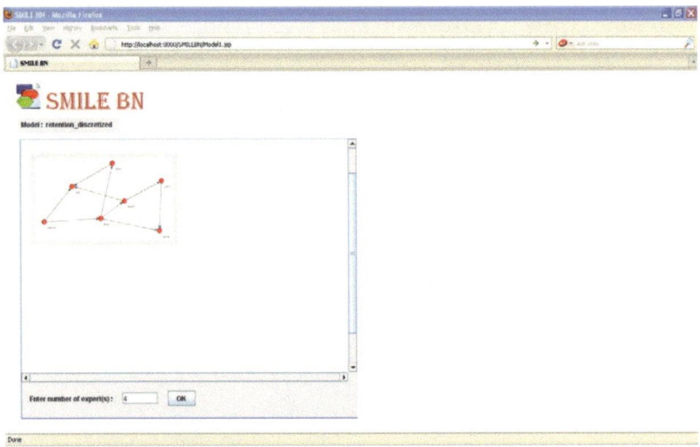

Fig. 9. Specifying the number of the expert(s)

We propose a methodology based on group decision making for weighting expert opinions or the degree of an expert's belief in identifying the causal relationships between variables in a BN model. The idea is to find the final BN solution that is obtained from a group of experts and to minimize the number of relationships among the nodes in the model for simplicity by setting a threshold value. The methodology consists of three sequential steps.

First, in a pre-processing step, all the experts in group must agree with each other for the BN model that is built based on the structure learning algorithms.

Second, we map every pair of causal variables into alternatives. Then, experts sort the alternatives by means of a fixed set of linguistic categories; each one has associated a numerical score. We average the scores obtained by each alternative and we consider the associated preference. Then we obtain a distance between each individual preference and the collective one through the Euclidean distance among the individual and collective scoring vectors. Taking into account these distances, we measure the agreement in each subset of experts, and a weight is assigned to each expert. We calculate the collective scores after we weight the opinions of the experts with the overall contributions to agreement. Those experts whose overall contribution to the agreement is negative are excluded and we re-calculate the decision procedure with only the opinions of the experts which positively contribute to agreement. The sequential decision procedure is repeated until it determines a final subset of experts where all of them positively contribute to agreement for group decision making. Lastly, we transform the alternatives and the collective scores that we obtain from previous step into the BN models. The mathematical formulas for this scheme are mentioned in section 4.

In the application point of view, users select the number of the experts (See Fig.9). In this example, we have a group of four experts who participate in identifying the degree of influential effects for the causal relationships in a BN model. The level of influential effects among the nodes based on each expert's belief is specified (See Fig.10). Each expert is asked to perform this task one by one. When all experts have completed this task, the BN model with the degree of expert's belief among causal relationship variables in the initial step of the decision procedure is presented (See Fig.11). Fig.12 shows the BN model and the degree of expert's belief among variables in normalized form (0..1) when users click on the "Normalized" button. Fig.13 shows the simplified BN model in the initial step of the decision procedure when users set a threshold value and click on "OK" button. They can select the other steps of the decision procedure from the list in a combo box below the model window and perform the same steps as presented in Fig.12 and Fig.13. The number of steps of the decision procedure depends on the number of expert and the ways they identify the degree of influential effects for the causal relationships in a BN model. Fig.14 - Fig.16 shows the BN

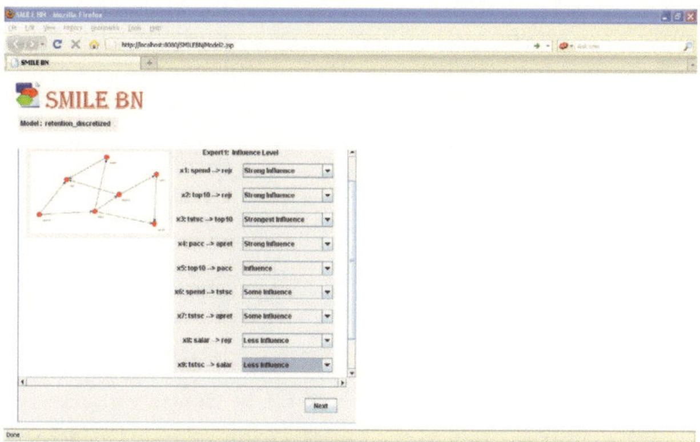

Fig. 10. Specifying the level of influential effects among the nodes based on expert's belief

model, the model in normalized form, and the model with a threshold value = 0.2 in the first step of the decision procedure. Fig.17 – Fig.19 shows the BN model, the model in normalized form, and the model with a threshold value = 0.3 in the second step of the decision procedure.

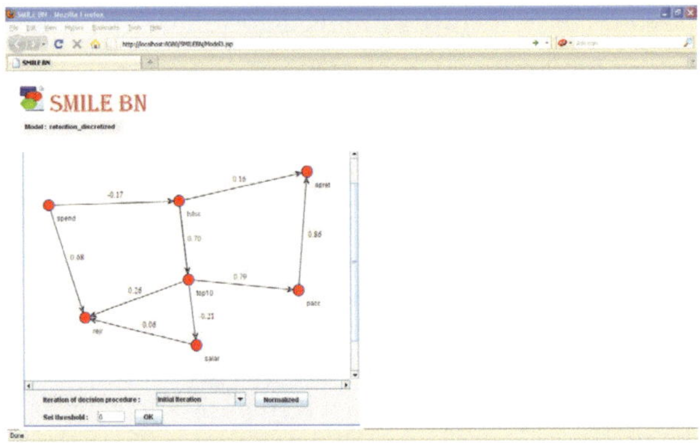

Fig. 11. BN model and the degree of expert's belief among causal relationship variables (initial step of the decision procedure)

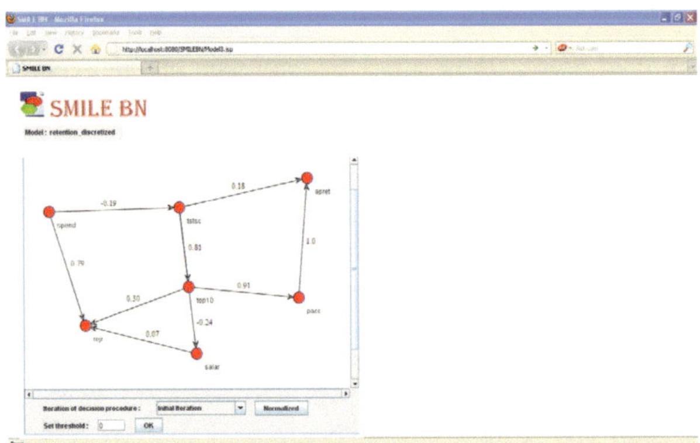

Fig. 12. BN model and the degree of expert's belief among causal relationship variables in normalized form (initial step of the decision procedure)

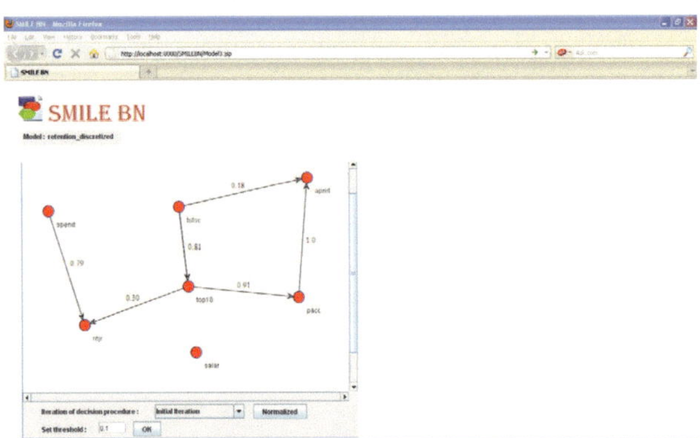

Fig. 13. BN model and the degree of expert's belief among causal relationship variables after applying the threshold value (initial step of the decision procedure)

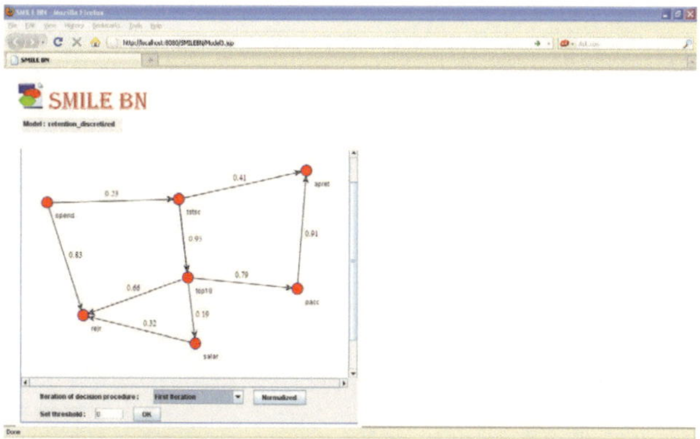

Fig. 14. BN model and the degree of expert's belief among causal relationship variables (first iteration of the decision procedure)

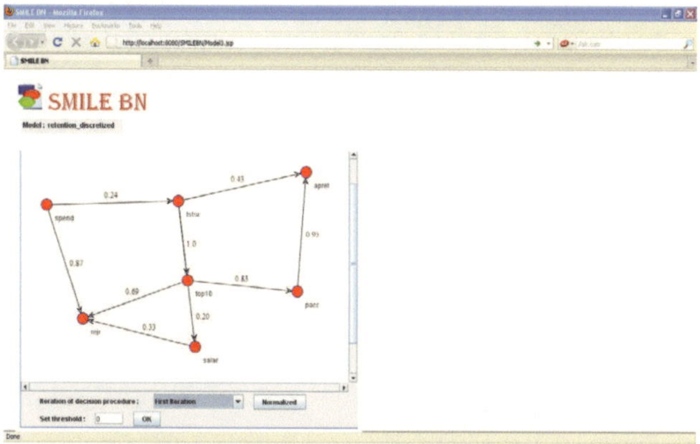

Fig. 15. BN model and the degree of expert's belief among causal relationship variables in normalized form (first iteration of the decision procedure)

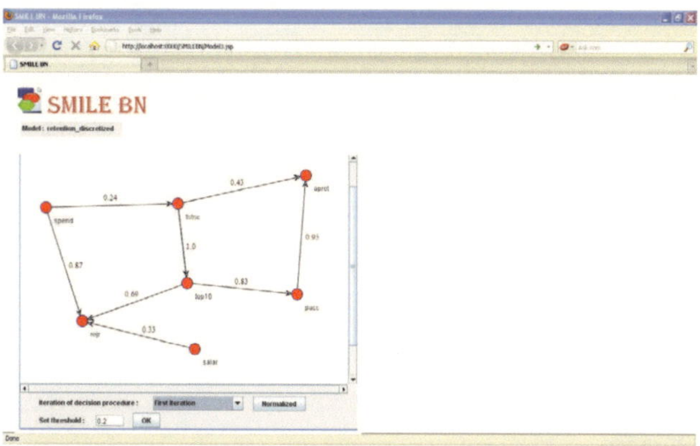

Fig. 16. BN model and the degree of expert's belief among causal relationship variables after applying the threshold value = 0.2 (first iteration of the decision procedure)

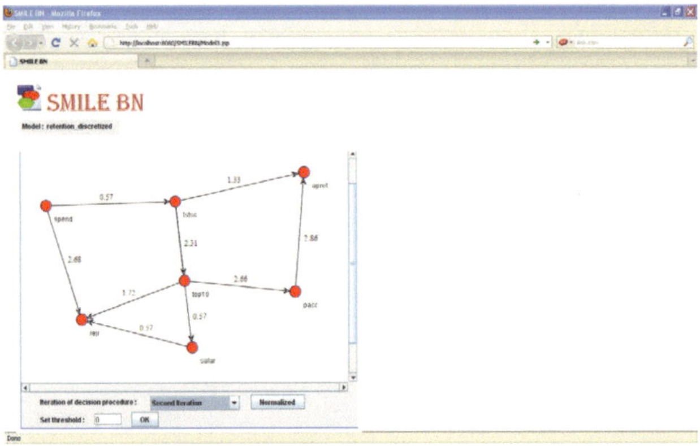

Fig. 17. BN model and the degree of expert's belief among causal relationship variables (second iteration of the decision procedure)

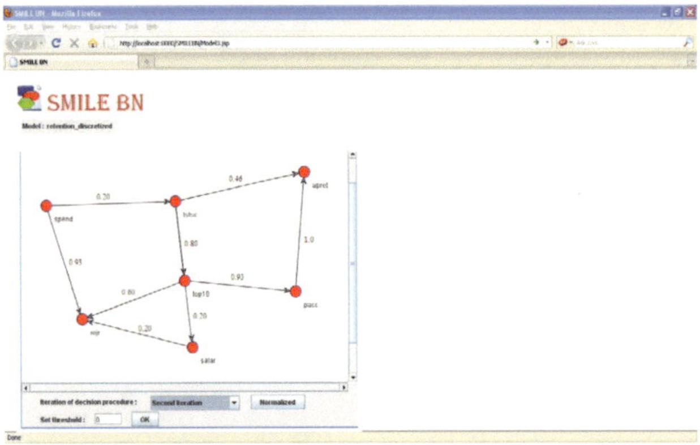

Fig. 18. BN model and the degree of expert's belief among causal relationship variables in normalized form (second iteration of the decision procedure)

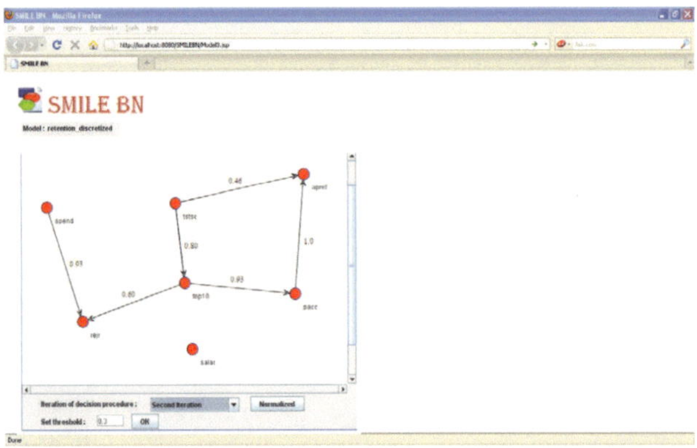

Fig. 19. BN model and the degree of expert's belief among causal relationship variables after applying the threshold value = 0.3 (first iteration of the decision procedure)

6. Conclusion and future work

This article presents a SMILEBN web application for building a Bayesian network model. The SMILEBN can build a BN model based on using two approaches. First, a BN model is built by applying the structure learning algorithms to a dataset. The variables in a dataset can be both discrete and continuous variables. The core reasoning engines of the SMILEBN web application consist of SMILE, SMILEarn, and JSMILE. SMILE is used for graphical probabilistic models and provides functionality to perform diagnosis. SMILEarn is used for obtaining data from a data source, pre-processing the data, and learning the causal structure of BN models. JSMILE is used for accessing the SMILE library from the web-based interface. Second, group decision making technique for weighting expert opinions scheme is applied to the BN model. This scheme is used to identify influential effects from parent variables to child variables in the BN model based on having information about the agreement and overall agreement measure produced by a group of experts. The sequential decision procedure is repeated until it determines a final subset of experts where all of them positively contribute to agreement for group decision making. Several steps of the decision procedure will be generated. The aims of the second approach are that we need to obtain the BN model, which all the experts agree to use, and to minimize the number of relationships among the nodes in the model for simplicity by setting a threshold value. When the number of relationships among the nodes decreases, the complexity of the conditional probability table on each child node also decreases.

Our future work will focus on improving a decision-oriented diagnosis approach. The SMILEBN will be extended to cope with influence or relevance diagrams.

7. Acknowledgement

The authors would like to thank the Decision Systems Laboratory, University of Pittsburgh for supporting documents, and source file of the engines: Structural Modeling, Inference, and Learning Engine (SMILE), SMILEarn, and JSMILE wrapper. All necessary files and documentations have been obtained from the Decision Systems Laboratory's web site. It is available at http://genie.sis.pitt.edu.

8. References

Bosch, R. (2005). Characterizations of voging rules and consensus measures. *Ph.D. Dissertation, Tilburg University*.

Cook, W. D.; Kress, M. & Seiford L. M. (1996). A general framework for distance-basedconsensus in ordinal ranking models. *European Journal of Operational Research*, pp. 392–397.

Druzdzel, M.J. (1999). SMILE: Structural Modeling, Inference, and Learning Engine and GeNIe: A Development Environment for Graphical Decision-Theoretic Models. *In Proceedings of the Sixteenth National Conference on Artificial Intelligence (AAAI–99)*, p. 902-903, Orlando, FL.

Herrera, F. & Herrera-Viedma, E. (2000). Linguistic Decision Analysis: Steps for Solving Decision Problems under Linguistic Information. *Fuzzy Sets and Systems 115*, pp. 67–82.

Jongsawat, N. & Premchaiswadi, W. (2009). A SMILE Web-Based Interface for Learning the Causal Structure and Performing a Diagnosis of a Bayesian Network. *Proceedings of IEEE on Systems, Man, and Cybernetic, Systems Science & Engineering (SMC2009)*, San Antonio, Texas, USA.

Jongsawat, N. & Premchaiswadi, W. (2010). Weighting Expert Opinions in Group Decision Making for the Influential Effects between Variables in a Bayesian Network Model. *Proceedings of IEEE on Systems, Man, and Cybernetic, Systems Science & Engineering (SMC2010)*, Istanbul, Turkey, pp.1029-1035.

Jongsawat, N.; Poompuang, P. & Premchaiswadi, W. (2010). Dynamic Data Feed to Bayesian Network Model and SMILE Web Application. *Bayesian Network edited by Dr.Ahmed Rabai*, p. 155-166.

Lacave, C.; Luque, M. and Díez, F.J. (2007). Explanation of Bayesian Networks and Influence Diagrams in Elvira. *IEEE Transactions on Systems, Man, AND Cybernetics-Part B: Cybernetics*, Vol. 37, No. 4, p. 952-965.

Mylltmaki, P.; Silander, T.; Tirri, H. & Uronen, P. (2002). B-Course a Web-based Tool for Bayesian and Causal Data Analysis. *International Journal on Artificial Intelligence Tools*, Vol. 11, No. 3, p. 369-387.

Poompuang, P.; Kungtasthan, A. ; Jongsawat, N. & Sutheebanjard, P. (2007). Graphical Decision-Theoretic Models on the Web. *Proceedings of Knowledge Management*, p. 163-170, Bangkok, Thailand.

Premchaiswadi, W. & Jongsawat, N. (2010). Bayesian Network Inference with Qualitative Expert Knowledge for Group Decision Making. *Proceedings of IEEE on on Intelligent Systems (IS 2010)*, London, UK, pp.126-131.

Tungkasthan, A.; Poompuang, P. & Premchaiswadi, W. (2008). SMILE Visualization with Flash Technologies, *Proceedings of IEEE on Software Engineering: Artificial Intelligence, Networking, and Parallel/Distributed Computing*, p. 551-556, Phuket, Bangkok, Thailand.

Yager, R. R. (1993). Non-Numeric Multi-Criteria Multi-Person Decision Making. *Journal of Group Decision and Negotiation 2*, pp. 81–93.

3

Using Dynamic Bayesian Networks for Investigating the Impacts of Extreme Events

Manoj K. Jha and Ronald A. Keele
Morgan State University
USA

1. Introduction

Investigating resiliency and interdependency of critical urban infrastructure has been the topic of interest in recent years (see for example, Zhang and Peeta 2011; Oh 2010). This is because of a surge in natural and man-made disasters over the last decade and limited resources available to cope with the resulting infrastructure failure. With an increased level of interdependencies among infrastructures, the potential for cascading failures are of a great concern. A cascading failure is one in which a failure in one infrastructure system causes the failure in one or more components of a second infrastructure (Rinaldi et al., 2001). Much of today's emergency preparedness research is heavily focused on what is considered by many to be the eight "critical" infrastructures: (1) Telecommunications, (2) Electric Power Systems, (3) Natural Gas and Oil, (4) Banking and Finance, (5) Transportation, (6) Water Supply Systems, (7) Government Services and (8) Emergency Services. Within the United States' transportation infrastructure systems, there exists approximately 5,000 public airports; 590,000 highway bridges; 120,000 miles of major railroad tracks; 2,000,000 miles of pipelines; 300 inland/coastal ports; 80,000 dams and 500 major urban public transit agencies. A large majority of these transportation infrastructure systems are highly interdependent with one another. The failure/collapse of one will more than likely cause the failure/collapse of another.

Urban infrastructure systems are vulnerable to a wide range of hazards from nature, technological errors, and human activities. Resiliency is connected with the recovery capacity of the infrastructure. These systems' interdependence introduce the added layer of uncertainty. Measures of resiliency are robustness, redundancy, resourcefulness and rapidity. Various classifications are used to define infrastructure interdependencies; however, classifications suggested by Rinaldi et al. (2001) are physical, geographic, cyber, and logical.

In this chapter we apply Dynamic Bayesian Networks (DBNs) for investigating resiliency and interdependency of critical urban infrastructure during extreme events. We study the decision framework for defining resiliency. We review different categories of modeling and performance measures of serviceability of the infrastructure in the face of extreme events.

2. Literature review

The word "resilience" is used in a variety of contexts and has been debated significantly since 1970. However, in light of terrorism threats and some natural disasters in the recent past, it is being studied in terms of the urban infrastructure. Resilient has also been connected with recovery capacity; for example, Primm (1984) suggested that it can be measured at the speed at which a system returns to its original state following an interruption.

Many approaches have been used to model infrastructure interaction including for example, agent-based models (Dudenhoeffer 2006), input-output models (Setola 2009), neural networks (Min and Duenas-Osorio, 2009) and scalable multi-graph methods (Svendsen and Wolthusen, 2007). As well as differing in their general approach, these methods differ widely in the type, size and number of networks being considered. The approaches can be combined in a collective model where different infrastructure networks are encompassed in a single model structure or a distributed type where each network is modeled separately and the results are passed between the models according to some mediating mechanism.

Agent-based models are computer simulations of systems where entities called agents are used to represent the behavior of system components. One notable example of agent-based modeling applied to the area of interdependent infrastructure is the Critical Infrastructure Modeling Software (CIMS) developed by a group at the Idaho National Laboratory (Dudenhoeffer et al., 2006). Input-output inoperability models (IIM) are financial models that have been used for analyzing the cascading effects in critical infrastructure systems (Setola et al., 2009). IIM uses inoperability levels to describe the state of each infrastructure network. A neural network is a collection of densely interconnected simple computing units called artificial neurons loosely based on the architecture of the human brain. Neural networks have been used for reliability analyses on interdependent infrastructures (Min and Duenas-Osorio et al., 2009). Scalable multi-graph models (Svendsen and Wolthusen et al., 2007) have been proposed as a means of representing both services that are consumed instantly (e.g., electricity and telecommunications) and those that exhibit buffering (e.g., water and gas) in the same model structure. The research group at the Idaho National Laboratory (INL) undertook a review of the state of the art in modeling critical infrastructure interdependencies in 2006. The group identified 30 modeling systems that could be applied to the interdependencies of critical infrastructure (Pederson 2006).

The Bayesian Network (BN) has recently become a popular method for coding uncertainty (see for example, Jha 2009). The use of BNs was proposed as an alternative approach to modeling the interdependencies of critical infrastructure (Buxton et al., 2010). Because an important feature of a BN model is the bidirectional reasoning that is a natural function of this model, it appears that modeling interdependent infrastructures works well with this concept. An infrastructure interdependency is a bidirectional relationship between two or more infrastructures through which the state of infrastructure A influences or is correlated to the state of infrastructure B, and vice versa (Grubesic and Murray, 2006).

Dynamic Bayesian Networks (DBNs) are extensions of BNs that take the time varying natures of various events into consideration (Jha 2009); thereby, allowing the modeling of close to real-world scenarios more realistically. This paper explores the use of DBNs for modeling transportation infrastructure interdependencies while considering the resiliency of impacted infrastructure.

3. Resiliency and interdependency of critical urban infrastructure during extreme events

The resiliency and interdependency of critical urban transportation infrastructure needs to be carefully explored during extreme events. The impacts of a particular hazard may be indirect because of the weaknesses in infrastructure systems. For example, in the event of earthquake, few properties are destroyed by the actual shaking but many are destroyed by fire. This example illustrates how an independent system of linked relationships connects a hazard event with its ultimate outcome (Little 2002). When analyzed separately, the impact of one disrupted infrastructure system can be fairly estimated; however, interdependence introduces an added layer of uncertainty. The nature of interdependence can be a cascading failure, where a disruption of one infrastructure causes disruption of another; escalating failure, where a disrupted infrastructure prohibits the recovery of another infrastructure that failed earlier; and common cause failure, where a disrupted infrastructure system fails as a result of a common cause such as a natural disaster.

Power systems are perhaps the most important component of critical infrastructure because other systems require a continuous flow of energy to operate. Communications and information infrastructure includes linkages which move data from point to point. This also is critical during emergencies. Transportation is an important component of the urban infrastructure which facilitates the flow of goods in and out of an urban area. Water and Wastewater systems in the cities are old and their upgrades are essential.

Within the literature, there are several concepts for measuring resiliency. The resilience triangle quantifies the loss of functionality from damage and disruption emerges from disaster research (Tierney and Bruneau, 2007). The resilience triangle helps to visualize the magnitude of the impacts of a disruption on the infrastructure. The depth of the triangle shows the severity of damage and the length of the triangle shows the time to recovery.

Resiliency of transportation infrastructure needs to be carefully and precisely investigated during extreme events. Given the uncertainty surrounding the hazard variables such as location, frequency and magnitude, we cannot anticipate and prevent all disasters. However, the reliability in the continuity of infrastructure systems can be ensured by countermeasures.

The R4 framework of resiliency (Bruneau et al., 2003) defined four measures for resiliency:

- Robustness, which is the ability of systems, system elements and other units of analysis to withstand disaster forces without significant degradation or loss of performance;
- Redundancy, which defines the extent to which systems, system elements, or other units are substitutable if significant degradation or loss of functionality occurs;
- Resourcefulness, which is the ability to diagnose and prioritize problems and to initiate solutions by identifying and mobilizing material, monetary, informational, technological, and human resources; and
- Rapidity, which is the capacity to restore functionality in a timely manner, containing losses and avoiding disruptions.

For transportation infrastructure, resiliency measures the availability of alternate routes, the reduction in total delay, the adaptive use of high occupancy vehicle lanes, and the ability to transfer passenger travel to other non-single occupancy vehicle modes to free up highway and roadway capacity to maintain freight mobility (Giuliano and Golob, 1998).

4. Bayesian networks as a decision-making tool

One of the main reasons for lack of coordination and poor decision-making in the face of an extreme event is the inability of comprehending the multitude of information, to maximize the utility of the single decision that needs to be made. For example, a quantitative measure of the reduced Quality of Life (QOL) due to forced migration in the wake of hurricanes may be difficult to estimate. The timing of well thought out decisions also plays a critical role since delay in decision-making by a split second may have devastating consequences. This is true with any critical situation, such as during a war which can be won or lost with one right or wrong decision. Extensive research using game theory has been done in this area.

Bayesian Networks (BNs) have been extensively applied in problems where causality, uncertainty, and interdependence among variables plays a role (Jha 2006 & 2009). Using a BN offers many advantages over traditional methods of determining causal relationships. Using BN, independence among variables is easy to recognize and isolate while conditional relationships are clearly delimited by a directed graph edge: two variables are independent if all the paths between them are blocked (given the edges are directional). Not all the joint probabilities need to be calculated to make a decision; extraneous branches and relationships can be ignored. The BN algorithm can run in linear time (based on the number of edges) instead of exponential time (based on the number of parameters). The theory of BN is available in standard references and only presented here briefly (Gamez et al., 2004; Jasen 2001).

Consider a domain U of n variables, $x_1,....,x_n$. Each variable may be discrete having a finite or countable number of states, or continuous. Given a subset X of variables x_i where $x_i \in$ U, if one can observe the state of every variable in X, then this observation is called an instance of X and is denoted as $X = p(x_i|x_1,...,x_{i-1},\xi)\vec{k}_X = p(x_i|\Pi_i,\xi)\vec{k}_X$ for the observations $x_i = k_i, x_i \in X$. The "joint space" of U is the set of all instances of U $= p(X = \vec{k}_X | Y = \vec{k}_Y)$, which denotes the "generalized probability density" so that $X = p(x_i|x_1,...,x_{i-1},\xi)\vec{k}_X = p(x_i|\Pi_i,\xi)\vec{k}_X$ given $Y = \vec{k}_Y$. For a person with current state information, $\xi\, p(X|Y,\xi)$ then denotes the "Generalized Probability Density Function" (gpdf) for X, given all possible observations of Y. The joint gpdf over U is the gpdf for U.

A Bayesian network for domain U represents a joint gpdf over U. This representation consists of a set of local conditional gpdfs combined with a set of conditional independence assertions that allow the construction of a global gpdf from the local gpdfs. One assumption imposed by Bayesian Network theory (and indirectly by the Product Rule of probability theory) is that each variable x_i, $\Pi_i \subseteq \{x_1,....,x_{i-1}\}$ must be a set of variables that renders x_i and $\{x_1,...x_{i-1}\}$ conditionally independent. In this way:

$$p(x_i|x_1,...,x_{i-1},\xi) = p(x_i|\Pi_i,\xi) \quad (1)$$

A Bayesian Network Structure then encodes the assertions of conditional independence in Eq. (1) above. Essentially then, a Bayesian Network Structure, B_s, is a directed acyclic graph such that: (1) each variable in U corresponds to a node in B_s, and (2) the parents of the node corresponding to x_i are the nodes corresponding to the variables in [Pi]$_i$. A Bayesian-network gpdf set Bp is the collection of local gpdfs $p(x_i|\Pi_i,\xi)$ for each node in the domain.

4.1 Handling uncertainty

Uncertainty is an attribute of information. A review of literature (e.g., see Klir 2002; Higashi and Klir, 1983) dealing with uncertainty reveals that a highly original, unorthodox theory of human affairs involving uncertainty was conceived and developed by George Shackle, a British economist and philosopher in early 1900. Shackle introduced the possibility theory to handle uncertainty.

4.1.1 Possibility theory

While probability theory has traditionally been used to handle uncertainty, in recent works (Klir 2002; Kikuchi and Chakroborty, 2006) use of possibility theory has been advocated, a theory originally pioneered by George Shackle. Kikuchi and Chakroborty (2006) note that the distinction between the two theories (possibility and probability) is rooted in the type of information they handle, and how it is formalized in a functional form, the distribution. The probability distribution represents much more specific (rigid) information than the possibility distribution. It is characterized by the concept of propensity, or actual occurrence of events. The additive property of the probability distribution clearly suggests consistency in the evidential support.

The possibility distribution, on the other hand, is founded on the concept of disposition, which implies "judgment" in the feeling of "possibility," "achievability," "acceptability," and "capacity of the events to occur." The possibility distribution covers a set of "possible ranges," less precise information than the probability distribution. *Hence, it is natural that how to express ignorance and uncertainty is an important part of the possibility theory framework.*

The possibility and necessity measures of possibility theory constitute the upper and lower bounds of probability measure. Conceptually, this is because only the possible events can be probable (Smets 1998). With a better quality of information, the difference between possibility and necessity measures narrows and each converges to the probability measure (Kikuchi and Chakroborty, 2006).

The value of probability is interpreted as propensity of occurrence of an event in an objective sense; and hence, it clearly has application to risk and uncertainty associated with strategic decision-making to seek countermeasures in the face of a possible attack by an adversary or hostile country. The value of possibility and necessity, on the other hand, is associated with the sense of force or momentum to support a particular decision alternative. Its uses are suited to comparing (ordering) two situations, or understanding the degree of uncertainty or degree of support for an alternative.

Uncertainty plays a key role in understanding the resiliency and interdependency of urban infrastructure during extreme events. In order to handle uncertainty, an integrated framework can be proposed in which the probability distribution of the Dynamic Bayesian Network (DBN) can be represented by a possibility distribution. In seeking decisions to go from one stage to the next, randomized decision rules can be implemented, similar to that proposed by Berger (1980).

4.2 Early model development

A year before the 9/11 attack on the World Trade Center and the Pentagon, a simulation model, called Site Profiler, using Bayesian Networks, had predicted that the Pentagon was a likely terrorist target. On that occasion, no one took the mathematical prediction seriously enough to do anything about it. The rest, as you know it, is HIS-TO-RY! Site Profiler (Hudson, Ware, Blackmond-Laskey and Mahoney, 2000), was developed after the bombing of U.S. Air Force servicemen in Khobar Towers, Saudi Arabia, in June 1996, in which 20 persons were killed and 372 were wounded, and the August 1998 bombings of the U.S. embassies in Dar es Salaam, Tanzania, and Nairobi, Kenya, where a total of 257 people were killed and more than 4,000 wounded.

The user of this system would enter information concerning a military installation's assets through a question-and-answer interface very similar to that of a tax preparation software package (Site Profiler actually modeled its interface on the one used in Turbo Tax).

Site Profiler was distributed to all U.S. military installations around the globe to assist the site commanders by providing the necessary tools to assess terrorist risks, to manage those risks, and to develop antiterrorism plans.

This synopsis should tell us two (2) things. First, is that mathematics can be a very powerful tool for assessing terrorist risks. Second, is that we need to think very carefully before discounting the results that the math produces, no matter how far-fetched many of them may seem.

4.3 Software

Modeling BNs can be a very difficult task. A number of commercial software packages are available for developing BBN based models. The more popular ones are (1) Analytica (Lumina, 2004); (2) Netica (Norsys, 2005); (3) Hugin (Hugin Expert A/S, 2004) and GeNie (DSL, 2005). Each package has its own strengths and weaknesses.

The Netica software is used to model the real-life examples presented later. Before constructing the Bayesian Network, a conceptual model of the scenario should be developed. The conceptual model will allow for conditional relationships to be developed prior to entering this information into the Netica software. The concept of conditional probability is very useful because there are numerous "real-world" examples where the probability of one event is conditional on the probability of a previous event.

5. Characterization of an extreme event

In the financial world, extreme events are termed "extraordinary items" which are defined as unusual in nature **AND** infrequent in its occurrence (Kieso, Weygandt and Warfield, 2007). Using this information, let us define an extreme event as an incident, that is; (a) unusual in nature **AND/OR** (b) infrequent in its occurrence. In our definition, both (a) and (b) do not have to take place simultaneously for an event to be classified as "extreme." Let us explore this matter in more depth.

Unusual in nature can be characterized as an event that possesses a high degree of uncertainty, such as a large magnitude earthquake occurring in Washington, D.C.

Infrequency of its occurrence can be characterized as an event that does not reasonably expect to occur in the foreseeable future, using the example above, an earthquake of a magnitude of 6.0 or greater occurring in Washington, D.C.

While many extreme events have occurred here in the United States, no better incident meets this definition than the terrorist attack on the World Trade Center on September 11, 2001. Many citizens living on the East Coast of the United States view the occurrence of a major earthquake as something that is not of an immediate concern. Many do not worry about it because the long held belief was that it WILL NEVER happen here!

Well it did! This extreme event occurred on Tuesday afternoon, August 23, 2011. At approximately 1:53 p.m., the strongest earthquake to hit the state of Virginia since May 31, 1897 took place. With a magnitude of 5.9, this rare earthquake rattled almost the entire East Coast, turning a lovely and calm Tuesday afternoon into one of total chaos. Cellular phone service was jammed, area buildings were evacuated and police/fire emergency dispatchers could not keep up with all of the incoming calls!

Earthquakes on the East Coast are rare, but they do happen, and these earthquakes are often concentrated in certain areas. One such area is the Central Virginia Seismic Zone, where the August 23, 2011 earthquake occurred. Unlike the state of California or the continent of Japan, Virginia is not located near the edge of a tectonic plate. Although the bedrock in this zone has no major faults, it is loaded with smaller faults that occurred when the Appalachian Mountains were formed.

Although it has been 114 years since a major earthquake of this magnitude has occurred, the August 23rd quake was a stark reminder that we can no longer assume the Alfred E. Neuman attitude of "What, Me Worry?" So, what can we do? Is it possible to model such events and reduce the inadequacy of our preparations and the great losses associated with these extreme events?

6. Example studies

In this section, we present several examples to investigate the resiliency and interdependency of critical infrastructure in extreme events. The first is the Virginia earthquake example whose analysis is presented without examining resiliency and interdependency. In the second and third example, a Dynamic Bayesian Network is employed to perform the analysis. The second example is from a power failure in a subway system operated by the Washington Metropolitan Area Transit Authority (WMATA). The third example is related to hurricane planning and preparedness.

6.1 Virginia's earthquake example

Can it happen again!? That was the question asked by many of the residents of the state of Virginia. Of course it can! But let us examine this in more depth. Since February 21, 1774, the state of Virginia has had only 20 (see Table 1) recorded earthquakes, ranging in magnitude from 1.9 (May 6, 2008) to 5.9 (May 31, 1897 & August 23, 2011).

Time	Number of Earthquakes
1700's	1
1800's	7
1900's	6
2000's	6

Table 1. Number of earthquakes in Virginia

Not shown in the above table is the fact that from 1774 to 1833 (59 years) there were no recorded earthquakes in the state of Virginia. Again, from 1975 to 2003 (28 years) there were no recorded earthquakes in the state of Virginia. This interval between earthquakes appears to have a rate of decay of 50 percent. It appears that the number of earthquakes in the state of Virginia is increasing. How!? Based on the table, it seems that the next earthquake will occur in the year of 2025. Using the year of 2011 as the benchmark and dividing 28 years by 2 and adding that number to the year 2011, we get the year 2025! But is that truly correct?

From 1774 to 2011 (237 years), there have only been 20 earthquakes in the state of Virginia. Given that information, there is an eight (8) percent [20/237] chance that the state of Virginia will experience an earthquake. That means there is a 92 percent chance that an earthquake will never occur. What about the earthquake's magnitude? Of the 20 earthquakes, only three (3) have had a magnitude greater than or equal to 5.0. That means there is only a 15 percent [3/20] chance of an earthquake having a magnitude of at least 5.0. There is an 85 percent chance that the earthquake will have a magnitude of less than 5.0.

Using the Bayesian method, let us try to answer the following question. What is the probability that the state of Virginia will have another earthquake with a magnitude of 5.0 or greater? Simple! Take the probability of Virginia experiencing an earthquake (0.08) multiplied by the probability of the earthquake having a magnitude of 5.0 or greater (0.15). To answer the question, there is a one (1) percent chance [0.08 x 0.15] that the state of Virginia will experience an earthquake having a magnitude of greater than or equal to 5.0. On the other hand, there is a 99 percent chance that the state of Virginia will experience an earthquake but with a magnitude of less than 5.0.

6.2 WMATA example

In order to illustrate this example, an artificial real-life scenario is constructed as follows: It is a clear and sunny Monday morning and you decide to take the WMATA metro subway system to work. You have an important 8:00 a.m. meeting and your boss is also attending. It is 7:10 a.m. and the train is moving from the Pentagon Station to Downtown Washington, D.C. via the "Yellow Line" (see Figure 1). In 10 minutes, you will be in the office. You will have enough time to get your coffee and to discuss the Washington Redskins victory over the Dallas Cowboys in yesterday's game.

Suddenly, the train abruptly stops! You hear the train operator say, "This train will be moving shortly!" Well, 15 minutes later, the train is still in the same position and the train operator again says, "This train will be moving shortly!" Suddenly, off go the lights! Everyone is in panic mode and you look at your cell phone and it is now 7:47 a.m.! You realize that you will not be making it to your meeting on time and you also notice that you have no cell phone service! You cannot even call the office and let them know where you are! An hour later you arrive in the office only to be met by the "steely" eyes of your boss and you decide it is not worth the trouble in trying to explain what in the heck happened.

The next morning you pick up the *Washington Post* and you read, "**Snake Cuts Power to Thousands of Pepco Customers**" (Hedgpeth, 2011). To your chagrin, the article states that five (5) circuit feeders were not working at a substation, leaving 6,800 customers without electricity and stranding several Metro trains. The article also quoted a Pepco spokesperson as saying, "The snake got stuck inside a breaker and was electrocuted!" Of course, your next reaction was, "How in the Sam Hill does something like this happen!?" "Aren't these people supposed to be prepared for anything!?"

Fig. 1. Map of the WMATA Rail System (courtsey of www.wmata.com)

Although the above story is fictional, the facts concerning the power failure are real. No Metro trains were stranded on that day. That information was only included to round out the story.

6.2.1 Constructing the Dynamic Bayesian network

A conceptual model of the above example is created as shown in Figure 2. Several extreme event scenarios are created using the WMATA example. First, if there is no power outage (see Figure 2), there is a 92.3% chance that the Metro trains and its passengers will **NOT** be stranded. But, a funny thing happened when the scenario was switched! If there is a power outage (see Figure 3), there is a 96% chance that the power outage is weather-related and there is a 92.3% chance that the Metro trains and its passengers will **BE** stranded.

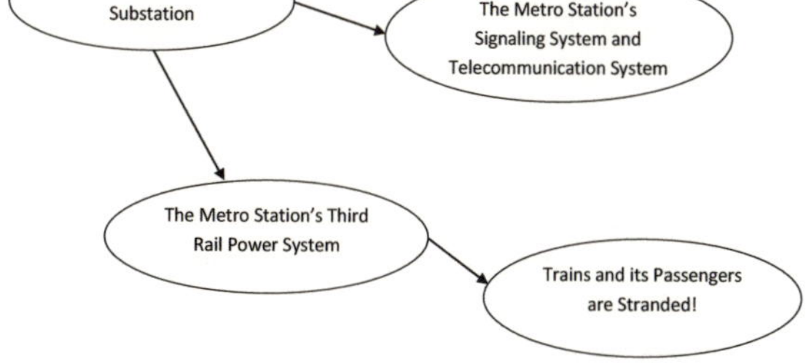

Fig. 2. A conceptual extreme event model for the WMATA example

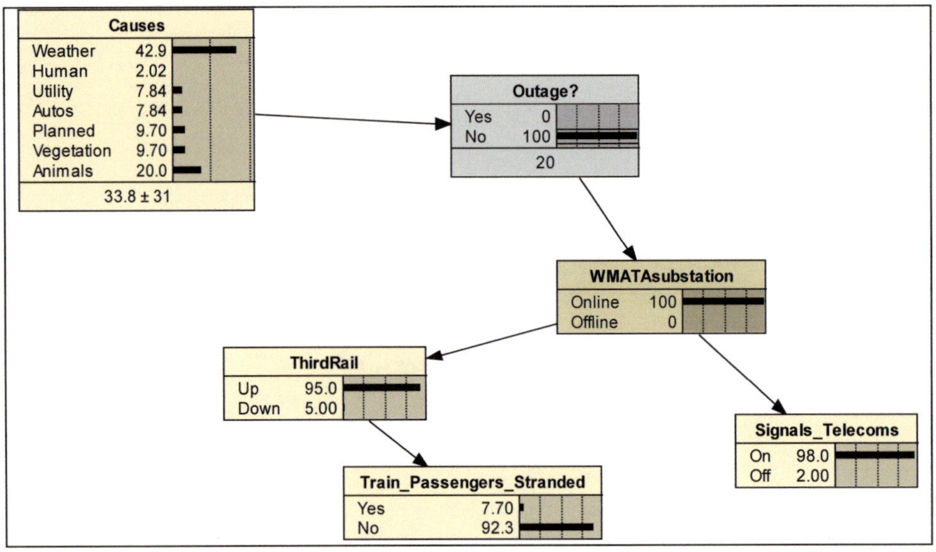

Fig. 3. A Dynamic Bayesian Network (DBN) with **NO** power outage

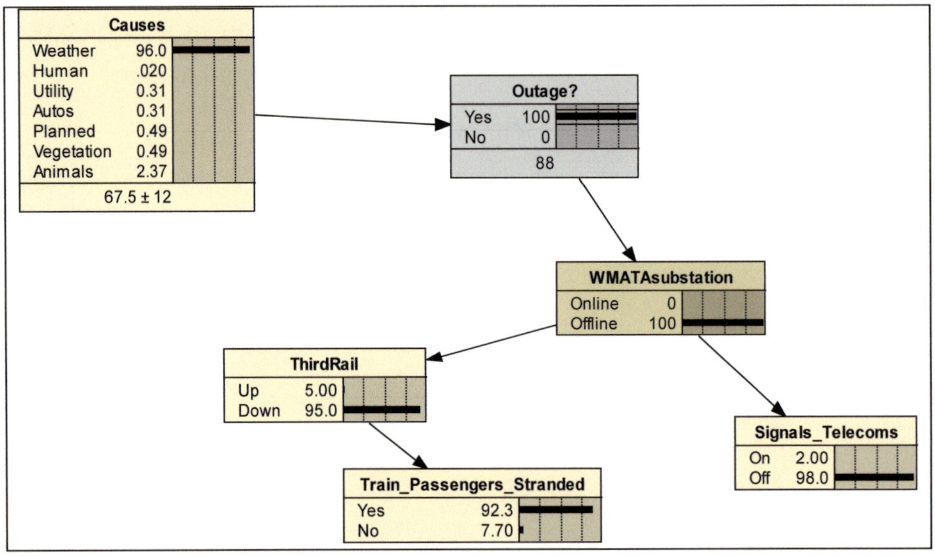

Fig. 4. A DBN with power outage

Although many approaches have been used to model infrastructure interdependencies, it was demonstrated, in principle, that DBNs can be used for modeling and evaluating interdependent infrastructures. In addition, the use of the Netica software allows for the modeling and evaluating of complex transportation infrastructure interdependencies. As we build newer systems, the complexity of these systems is steadily increasing and becoming more and more interdependent. Also, the operation of these systems is so complex that it defies the understanding of all but a few experts, and sometimes even they have incomplete information about the system's potential behavior. But, with the use of DBNs, modeling these complexities should become much easier in the future.

6.3 An example of hurricane planning and preparedness with a DBN

In the case of an impending hurricane a DBN can be used to plan for evacuation and displacements based on a threshold value of the probability of the extent of the disaster. Consider the case of hurricane Katrina which had multiple decision-makers with multiple perceptions of the impending disaster. Moreover, in the case of Katrina it was not clear who had the authority to order evacuations as necessary. Weather prediction centers, such as the hurricane center in Miami generally do a very good job in plotting the path of the impending hurricane and its severity. We can attach a probability of severity due to an impending hurricane using the weather predictions over a planning horizon.

An application of a Dynamic Bayesian Network for predicting the Quality-of-Life (QOL) of displaced citizens due to a hurricane, such as Katrina is shown in Figure 5. It represents a Directed Acyclic Graph (DAG). Figure 5 legends are shown in Table 2. If comprehensive

Legend	Explanation
Weather Forecast	Time-dependent weather forecast that predicts the category of a hurricane and its path
Contingency Measures	Contingency measures in place in the wake of an impending hurricane
Decision Maker's Action	Measure of coordinated response of decision-makers in the face of an impending hurricane
Damage Severity	Damage caused by a hurricane, measured in three categories: low, medium, high
Quality-of-Life Measure	Extent of degradation in the quality-of-life of displaced population

Table 2. Figure 5 Legend

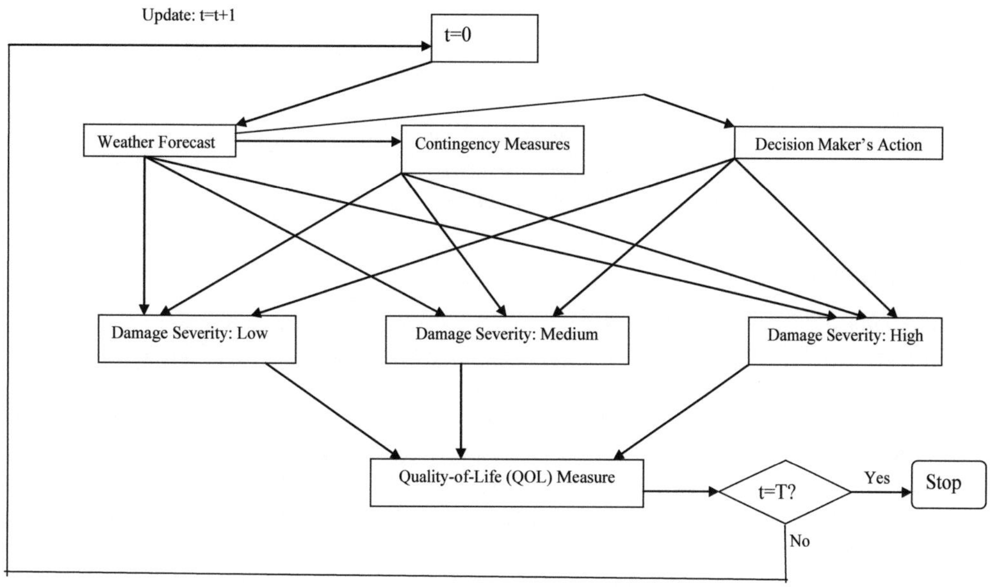

Fig. 5. A DBN Application for Hurricane Evacuation and Planning

data were available the DBN shown in Figure 5 is capable of estimating the QOL of the displaced population and also relative effectiveness of contingency measures and decision maker's actions. A sensitive analysis can also be conducted if additionl data were available. Having such a tool will allow decision-makers to take timely and coordinated measures in the wake of an impending hurricane to minimize the degradation of the QOL of displaced population.

7. Conclusions and future works

In this Chapter, we discussed the resiliency and interdependency of critical urban infrastructure systems in extreme events, and showed the applicability of Dynamic Bayesian Networks (DBNs) in examining resiliency and interdependency of such systems through a series of examples. The key contribution of our work lies in the critical analysis of extreme events and their impact on urban infrastructure systems and recognizing DBN as a valuable tool to model resiliency and interdependency.

In future works, a planning model can be developed using the DBN and a simulation tool, for a robust and sustainable community hard hit by catastrophic natural disasters. Such a model can:

- Measure the relative vulnerability of different geographic regions of the world to some key natural hazards, such as earthquakes, hurricanes, and floods.
- Identify development factors that contribute to risk, and show in quantitative terms, how the effects of disasters could be either reduced or exacerbated by policy choices.
- Demonstrate the ways in which development contributes to the configuration of risk and vulnerability.
- Provide quantitative evidence to advocate for the reorientation of development policy and planning in a way that contributes to the management and reduction of disaster risk.

8. Acknowledgment

This work was completed at the Center for Advanced Transportation and Infrastructure Engineering Research (CATIER) at the Morgan State University. It was partially funded by the Department of Energy-Massie Chair of Excellence Program.

9. References

Baltimore Gas and Electric Company 2011. What Causes Power Outages? www.bge.com.

Berger, J. (1980). Statistical Decision Theory: Foundation, Concepts, and Methods, Springer-Verlag.

Bowers, S. 2004. Graphical Probability Models for Bridge Management. A Research Paper submitted to the NSF-REU of the University of Delaware.

Bruneau, M., S.E. Chang, R.T. Eguchi, G. C. Lee, T.D. O'Rourke, A.M. Reinhorn, M. Shinozuka, K. Tierney, W.A. Wallace, and D. von Winterfeldt (2003). "A Framework to Quantitatively Access and Enhance the Seismic Resilience of Communities". Earthquake Spectra, Vol. 9, No. 4, pp. 733-752.

Buxton, R., Uma, S.R., & King, A.B. 2010. Modeling Interdependences of Critical Infrastructure. *Proceedings of the 2010 NZSEE Conference.*

Charnaik, E., 1991. Bayesian Networks without Tears. *AI Magazine*, Vol.12, No. 4, 50-63.

Dudenhoeffer, D.D., Permann, M.R. & Manic, M. 2006. CIMS: A Framework for Infrastructure Interdependency Modeling and Analysis. *Proceedings of the 2006 Winter Simulation Conference.*

Gamez, J.A., Moral, S., Salmeron, A. (eds.) (2004), Advances in Bayesian Networks, Springer-Verlag, New York.

Giuliano, G., and J. Golob (1998). "Impacts on Northridge Earthquake on Transit and Highway Use" Journal of Transportation Statistics, 1(2):1-20. May.

Greenberg, R., Cook, S.C. & Harris, D. 2005. A Civil Aviation Safety Assessment Model Using a Bayesian Belief Network (BBN). *The Aeronautical Journal*, 557-568.

Grubesic, T. & Murray, A.T. 2006. Vital Nodes, Interconnected Infrastructures and the Geographies of Network Survivability.

Hedgpeth, D. 2011. "Snake Cuts Power to Thousands of Pepco Customers." *The Washington Post*, B6.

Higashi, M. and Klir, G.J. (1983). Measures of Uncertainty and Information Based on Possibility Distributions, Int. Jrl. Of General Systems, 9(1), 43-58.

Hudson, L.D., Ware, B.S., Blackmond-Laskey, K., Mahoney, S.M. 2000. An Application of Bayesian Networks to Antiterrorism Risk Management for Military Planners.

Infrastructure Canada-Background Paper on Resilient Cities (March 2004)

Jasen, F.V. (2001), Bayesian Networks and Decision Graphs, Springer-Verlag, New York.

Jha, M.K. (2006). Applying Bayesian Networks to Assess Vulnerability of Critical Transportation Infrastructure, in *Applications of Advanced technology in Transportation*, pp. 1-5, Wang, K.C. et al. (eds.), ASCE Press, Reston, VA.

Jha, M.K. (2009). A Dynamic Bayesian Network for Predicting the Likelihood of a Terrorist Attack at Critical Transportation Infrastructure Facilities, *Journal of Infrastructure Systems*, 15(1), 31-39.

Jha, M.K. (2010). Optimal Highway Infrastructure Maintenance Scheduling Considering Deterministic and Stochastic Aspects of Deterioration, in *Sustainable and Resilient Critical Infrastructure Systems*, K. Gopalakrishnan and S. Peeta (eds.), 231-248, Springer.

Kieso, D.E., Weygandt, J.J. and Warfield, T.D. (2007). Income Statement and Related Information. *Intermediate Accounting, 12th edition*, 137.

Kikuchi, S. and Chakroborty, P. (2006). Place of Possibility Theory in Transportation Analysis, Transportation Research-B, 40, 595-615.

Klir, G.J. (2002). Uncertainty in Economics: The Heritage of G.L.S. Shackle, Fuzzy Economic Review, VII(2), 3-21, Nov. 2002.

Lee, E.E., Mitchell, J.E., and Wallace, W.A. (2004). Assessing Vulnerability of Proposed Designs for Interdependent Infrastructure Systems. In Proceeding of 37th Hawaii International Conference on System Science, IEEE, 54-61.

Little, Richard G. (2002) Toward More Robust Infrastructure: Observations on Improving the Resilience and Reliability of Critical Systems. Proceedings of 36th International Conference on System Science: University of Hawaii at Manoa.

Marcot, B.G., 2005. What are Bayesian Belief Network Models?

Min, X. & Duenas-Osorio, L. 2009. Inverse Reliability-Based Design of Interdependent Lifeline Systems. *TCLEE 2009: Lifeline Earthquake Engineering in a Multi-hazard Environment.*

Montironi, R., Whimster, W.F., Collan, Y., Hamilton, P.W., Thompson, D. & Bartels, P.H. 1996. How to Develop and Use a Bayesian Belief Network. *Journal of Clinical Pathology*, Vol. 49, 194-201.

Netica, Norsys Software Corporation, www.norsys.com.

Oh, E. H. (2010). Impact Analysis of Natural Disasters on Critical Infrastructure, Associated Industries, and Communities, Ph.D. Thesis, Purdue University.

Pederson, P., Dudenhoeffer, D., Hartley, S. & Permann, M.R. 2006. Critical nfrastructure Interdependency Modeling: A Survey of U.S. and International Research. Report prepared by the Idaho National Laboratory.

Primm, Stuart L. (1984). The Complexity and Stability of Ecosystems" Nature 307:321-326

Rinaldi, S.M., Peerenboom, J.P. & Kelly, T.K. 2001. Identifying, Understanding and Analyzing Critical Infrastructure Interdependencies. *IEEE Control Systems Magazine*, Vol. 21, No. 6, 11-25.

Setola, R., De Porcellinis, S. & Sforna, M. 2009. Critical Infrastructure Dependency Assessment Using the Input-Output Inoperability Model. *International Journal of Critical Infrastructure Protection*, Vol. 2, 170-178.

Sheffi, Y and J. Rice (2005)." A Supply Chain View of Resilient Enterprise", Sloan Management Review, 47 (1):41-48

Smets, P. Theories of Uncertainty. In Ruspini, R.E., Bonissone, P.P., Pedrycz, W. (eds.), Handbook of Fuzzy Computation. Institute of Physics Publishing, Bristol and Philadelphia, pp. 1.2.1-1.2.9.

Svendsen, N.K. & Wolthusen, S.D. 2007. Connectivity Models of Interdependency in Mixed-Type Critical Infrastructure Networks. *Information Security Technical Report*, Vol. 12, 44-55.

Tierney, K. and M. Brunuau. (2007). "Conceptualizing and Measuring Resilience: A Key to Disaster Loss Reduction". TR News 2006. Vol. 250, No.14, 2007, pp. 15.

Vurgin, E Warren, et al.; Framework of Assessing the Resilience of Infrastructure and Economic System- Sustainable Resilient Critical Infrastructure Systems – Springer 2010.

Zhang, P. and S. Peeta (2011). A Generalized Modeling Framework to Analyze Interdependencies among Infrastructure Systems, Transportation Research, Part B, 45(3), 553-579.

4

A Spatio-Temporal Bayesian Network for Adaptive Risk Management in Territorial Emergency Response Operations

Alberto Giretti, Alessandro Carbonari and Berardo Naticchia
Università Politecnica delle Marche, DICEA Department,
Building Construction Team Ancona
Italy

1. Introduction

In this Chapter we intend to show that Bayesian Networks may act as an excellent Decision Support System (DSS), even when very complex risk scenarios must be evaluated in real time.

The complexity of the specific application described derives not only from risk variability depending on spatial and temporal domains, but also from the high number of variables involved.

Forest Fire Management is a tough task for emergency squads, because they are called on to take decisions very quickly and control vast territorial areas. In addition, the behaviour of fire is strongly dependent on weather and soil conditions, and is hence difficult to predict in a short time through the concurrent implementation of consolidated analytical models.

For this reason the availability of a Bayesian model is the only solution capable of allowing emergency squads to exploit the backward propagation of Bayesian networks and perform real-time diagnoses. In other words, once the user sets the desired state for the output variable(s), the network is able to work out the most likely state values for the corresponding input, thereby helping to discern how to act optimally in order to control the spread of fire. The same tool may also be used for scenario analyses.

The main problems tackled in this chapter are connected with the spatio-temporal nature of the spread of forest fire and with the complexity caused by the great number of time slices involved, which are proportional to the duration of the phenomenon.

The suggested solution is to model the spatial features through the use of Object Oriented Bayesian Networks, where every elementary territorial area is mapped over one of the elementary pieces of the whole network. To that end, the evolution of phenomena over time was modelled through the approach known as Dynamic Bayesian Networks: the time step to complete any passage from one network to another is dictated by the time it takes for the fire to burn down any elementary territorial area and to move into any of the others. Due to the unknown number of needed time steps (because it is not known *a priori* how long any fire will last), the whole model was developed as a combination of dynamic networks and

VBA programmed tools. Programming was mainly needed to cope with the great number of variables involved and to find a computationally acceptable approach to spatio-temporal simulation, when there is no pre-determined limitation on the timeline.

The next paragraph will give some insight into the state-of-the-art regarding spatio-temporal Bayesian modelling, forest fire risk management and available techniques for CPT learning.

The third paragraph supplies the basic technical concepts, relative to forest fires (e.g. ranking, qualitative behaviour, variables involved in the phenomenon, analytical models presently used for their simulation etc..), which are relevant for the development of the desired Bayesian model.

The same paragraph also addresses how to build the graphical model, as well as how to estimate the quantitative strength of the connections between variables expressed as conditional probability distributions, how to translate the dynamic nature of the physical phenomenon into Bayesian formalism and how to overcome the barrier of NP-hard complexity for problems of this kind.

The fourth paragraph implements the quantitative part of the final model, including CPT estimation.

The fifth paragraph of this chapter shows how the Bayesian model developed in the previous parts can be translated into Visual Basic programme language and then interfaced with a Graphic User Interface (GUI). Finally, a validation of the whole model applied to a real case study in the forest of the Esino-Frasassi (Ancona, Italy) mountainous district will be proposed.

2. Scientific background

Bayesian Networks have been extensively celebrated for their unique capability to provide, at the same time, both intuitive and scientifically rigorous representations of complex systems. In addition, after validation, they can be used for performing both scenario analyses, through inference propogation algorithms, and diagnostic reasoning, through backward propagation based on the well known inversion rule (Pearl, 1988).

These networks also have the advantage of enabling qualitative and explicit representation, where nodes represent variables and arcs represent quantitative relationships among the same, worked out through parametric probabilistic models. We invite readers to refer to the numerous and well written reference texts available on the subject, such as (Korb & Nicholson, 2004; Jensen, 1996), for basic rules about how to develop robust models. In the rest of this paragraph we prefer to go into detail regarding spatio-temporal Bayesian networks and the Conditional Probability Table (CPT) estimation procedures used in the application presented here.

When the domains to be modelled are very complex, Object Oriented Bayesian Networks (OOBN) are usually used: they are made up of several elementary networks, sharing some of the variables, which constitute the links between the networks (Naticchia et al., 2007). Each elementary network is generally developed separately (and models one of the many physical phenomena involved) but the inference algorithms are propagated over the whole set of elementary networks.

A straightforward extension of this approach is given by Dynamic Bayesian Networks (DBN): these are based on a discretized time course and are made up of several time slices, each representing the state of the system at a particular moment in time. In this case, some of the variables have no fixed states, but change over time. Therefore connections hold not only between those variables linked by a causal relationship but also between the same variables represented in different time slices, because this takes into account their variability as time elapses.

These concepts are better explained in 2.1, which reports the algorithms used to implement CPTs in the networks developed for the Bayesian model for forest fire risk management.

Finally, paragraph 2.2 gives a brief survey of the procedures currently adopted to cope with forest fire risk management and the approaches used for operations in the event of emergencies.

2.1 Spatio-temporal Bayesian modelling

Spatial Bayesian Networks (SBN) are BNs that represent data regarding spatial domains and Spatial Dynamic Bayesian Networks (SDBN) are BNs that represent spatio-temporal data, that is spatial status changes over time. The application of BNs to model the evolution of processes that have temporal dynamics requires, in its simplest formulation (Neapolitan, 2004):

- an initial instance of the Bayesian network that contains the formulation of the problem at time t=0, that is the set of random variables $X_{i,0}$ and the related conditional probability distributions: $P(X_{i,0} | X_{i-1,0})$, $P(X_{i-1,0} | X_{i-2,0})$, etc.;
- one or more transition networks that correlate the variables of the BN instance at t=0 with the variables of the BN instance at t=1.

Fig. 1 shows a graphical representation of three time slices of a DBN.

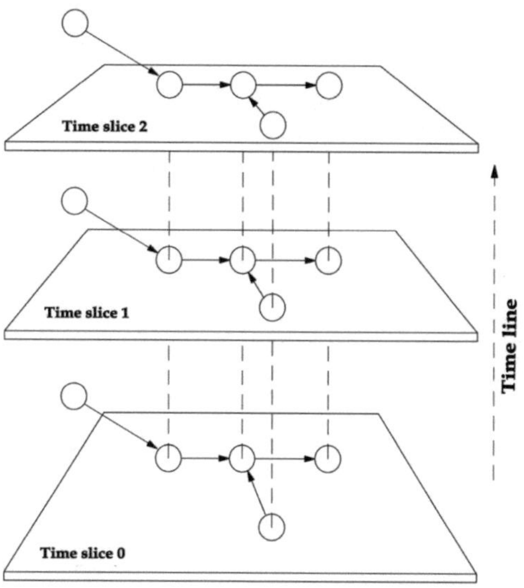

Fig. 1. Graphical representation of a Dynamic Bayesian Network: it is made up of three instances of the same BN. Yellow dashed lines represent the transition network. The evidence variables are represented by the nodes placed outside the time plane.

Two simplifying assumptions can usually be made about the physical processes at hand:

- all the information needed to predict the state of the process at time t+1 is contained in the description of the process state at time t. No information about earlier time is needed. These kinds of processes are called Markov processes of order one;
- the process is steady, that is, the transition networks remain the same for any $t_i \rightarrow t_{i+1}$.

The representation of the spatial evolution of a dynamic process by means of SBNs requires, above all, the domain space to be tessellated in such a way that each tessera represents a portion of the space with uniform behaviour. We will call this portion of space a cell. If the overall domain space is relatively uniform, different instances of the same type of BN can be used to represent different cells. In our case for example, the whole territory is covered by different types of fuel loading (e.g. grass, conifer trees etc.) which have essentially the same fire dynamics. Therefore each cell is represented by means of a different instance of the same BN. Each instance is then specialised for its fuel-loading type by means of a set of parameters that are modelled as evidence variables. Secondly, in order to grant the spatial continuity of the process evolution, the transition network must involve only neighbour cells. In our case this means, for example, that a cell cannot be ignited if neither of its neigbours is (please refere to Fig. 2).

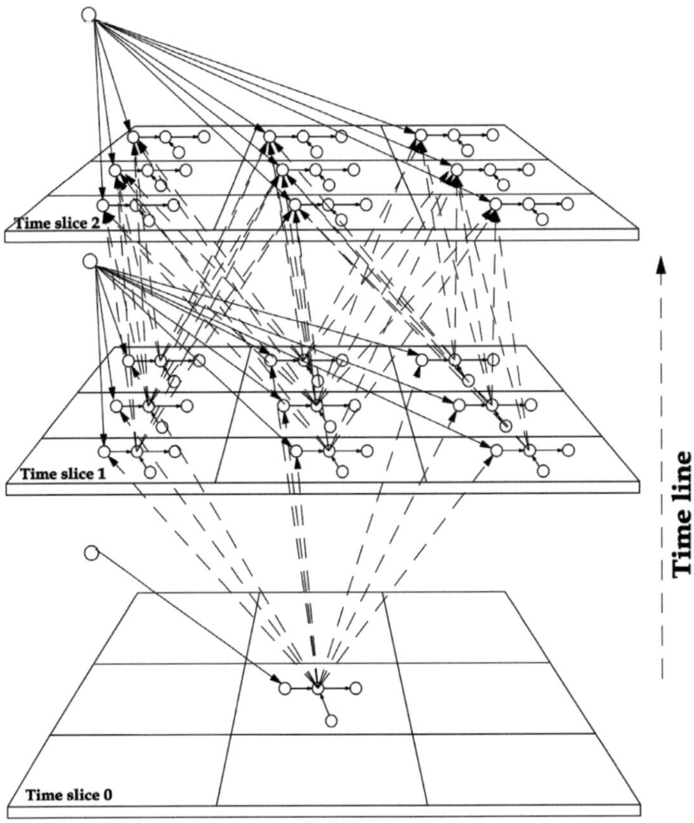

Fig. 2. Graphical representation of a Spatial Dynamic Bayesian Network, where the DBN is made up of three instances of the same BN. Dashed lines represent the transition network.

The evidence variables are represented by the nodes placed outside the time plane. Due to the space continuity assumption, transitions between different time slices occur only through neighbour cells.

Modelling spatio-temporal processes through SDBNs does not differ in principle from the standard methods of BN modelling, therefore it requires the definition of the cell structure, of the transition network structure and the implementation of the CPTs. The methodologies for direct implementation of network structures (both cells and transition) through domain knowledge modelling, using for example techniques like parent divorcing, temporal transformations, unidirected dependence relations, etc. (Kjaerulff & Madsen, 2008), are still applicable, as well as Data-Driven Modelling, like PC and NPC algorithms.

The problem arising with SDBN implementation is mostly related to complexity. Data-driven structure identification of SBDN is usually hindered by the great number of available options. Some algorithms are currently being developed to drive the optimal structure selection (Tucker & Liu, 2004).

In the case of a large SBDN, standard procedures suggest a two step approach: first learning the structure of the cell network, and then, after constraining the identified cell structure, learning the transition network. This usually helps in keeping the complexity to a manageable level.

EM algorithms provide batch CPT parameters for learning capabilities from data. If data are not available direct implementation of CPTs from domain data is the other option. In our case we found the application of a simple Montecarlo simulation tool very useful; this is implemented in many commercial BN packages, that are able, through a Montecarlo simulation, to map an analytical equation onto CPT involving random variables with numerical interval domains (Hugin Expert, 2008).

A final remark should be made to highlight the computational complexity of SDBN. Probabilistic inference in an SDBN can be performed using standard algorithms. However, since the size of an SDBN can become enormous when the simulation continues for a long time and when the domain is large, the algorithms may be quite inefficient and/or the network footprint in the computer memory may become impractical. There is a special subclass of SDBNs in which inference can be carried out more efficiently. This subclass includes BN in which the cell networks in different time steps are connected only through non-evidence variables. In that case, to update the probability of the current time step, we need only the values computed in the previous time step and the evidence at the current time step. This means that it is possible to implement an algorithm that keeps only the bare minimum network structure needed in order to represent two time steps (Neapolitan, 2004). In our case, excluding the cell "woodtype" and the "environmental" and "weather" variables from the transition network allowed us to implement an algorithm, as described in section 5, that inherited this necessary property.

2.2 Forest fire risk management

(Luke & McArthur, 1978) performed a lot of analyses regarding the spread of the fire frontline because this is affected by several parameters, such as environmental temperature, brush, topography and so on, which will be further detailed in paragraph 3.

Forest fires are usually ranked according to their types and are described according to evolution phases. (Brown & Davis, 1973) name the three basic types of fires according to the vegetation layer in which the fire is burning, that is ground, surface and crown fire:

- Ground fires: they spread in subsurface organic fuels, such as duff layers under forest stands, Arctic tundra or taiga, and the organic soils of swamps or bogs;
- Surface fires: they spread by flaming combustion through fuels on or near the surface – grass, shrubs, dead and downed limbs, forest needle and leaf litter, or debris from harvesting or land clearing;
- Crown fires: they burn through the tree crowns, they are often dependent on surface fires and are invariably ignited by surface fires.

The phenomenon generally evolves according to four burning phases:

- Initial build-up: when the fire's intensity, even if capable of keeping itself alive, is not capable of raising the temperature of the fuel, producing weak burning which is faster over earth covered with grass and low trees, than in high-tree forests;
- Transition stage: the fire frontline has increased to such an extent that it is able to dry the fuel (e.g. underwood and trees) and favour the spread, of fire also due to the flame angle and height which is affected by winds;
- Final stage: when the fire reaches its maximum intensity and strength, generating spotting phenomena, setting a balance with the external climate; in this phase external factors are less liable to influence the spread of fire which goes ahead autonomously;
- Extinction stage: the intensity of the fire decreases and its behaviour is newly determined by external factors (environmental and of the context in general), until extinction.

The measuring of intensity is of basic importance and one of the meaningful ranking approaches is known as Byram intensity (I), i.e. the product between the amount of heat generated per metre and the fire frontline propagation speed:

$$I = 0.007 \cdot H \cdot W \cdot R \tag{1}$$

where H is a parameter called heat of ignition (cal/g), W is the fuel weight per unit of surface (t/ha) and R is the propagation speed (m/min).

The greater the energy per unit of surface that is released by the fire, the greater is the damage caused to vegetation. However the frontline speed is another factor strongly affecting the difficulty involved in fighting the frontline. (Andrews & Rothermel, 1982) have worked out the "Fire Behaviour Characteristics Chart", where propagation speed is proposed as a function of the heat released per unit of frontline length and flame intensity per unit of surface (Fig. 3-a).

Of course there are a number of factors to be evaluated in order to determine the best counteraction for the spread of fire during emergency operations. Experts' decisions are generally based on a consolidated knowledge of the forest and on the symptoms indicating how the fire is behaving. One example is "blow-up", which is any quick evolution from transition to the final stage, suggesting that the fire will be very intense and generally caused by a mixture of high speed and intensity, feeding each other. Data about the humidity content of the fuel, topography and weather forecast are also of great help in supporting decisions (Scott & Burgan, 2005). A typical phenomenon encountered in high intense fire frontlines is known as spotting: organic matter is thrown into the air and floats as a result of the convective movement caused by the fire's intensity, until the wind scatters it and causes other fires to be triggered in the forest.

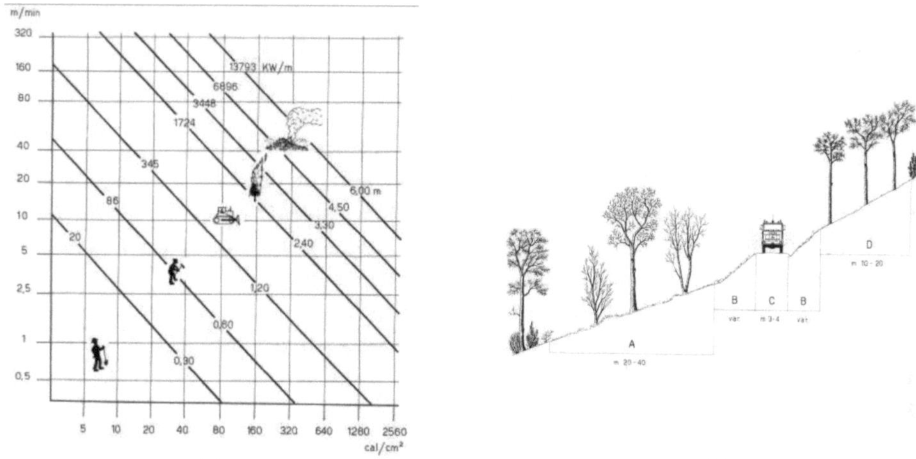

(a) (b)

Fig. 3. Fire Behaviour Characteristics Chart (a) and one example of preventive action (b)

All these and many more data must be evaluated when planning defensive action against the spread of fire. The general approach is to simulate fire behaviour (speed and intensity during propagation) under several boundary conditions, in order to design remedial action. The manager of emergency squads then evaluates the available input and decides on the best action to take. The availability of accurate simulators is critical during the early phase in the spread of fire because emergency action should be taken at this stage, before the fire burns up. Some of the most widely used simulators are: FOCUS™ (Fire Operational Characteristics Using Simulation), and FIRESCOPE™ (Fire-fighting Resources of Southern California Organized for Potential Emergencies). They are based on fire spread models like the ones described in the next paragraph and they drive the choice for the preventive or active response, to be adopted when the fire is burning.

The most usual preventive action is aimed at diminishing the amount of fuel available for feeding the fire (Brown & Davis, 1973): fuel is removed in specific areas, sometimes along strips of forest, or roads may be built in order to impede ignition or break the spread of the fire in the early phases (Fig. 3-b). However the efficiency of this type of response is limited when fires are very intense, hence active means are needed. The same authors also suggest drawing forest maps, where all the parameters are graphically represented and using them as reference for fire management.

Active means are ranked according to direct or indirect action: the former requires water or chemical products to throw over the fire, while the latter tries to clean the areas along the boundaries of the fire frontlines so as to hamper spreading.

It is clear that in any case, for both preventive action and active response, it is necessary to forecast how fires will behave.

The strategy suggested in this paper, and feasible once dedicated spatio-temporal Bayesian Networks are available, is to build an expert system containing the knowledge deriving from various fire models, which can support decision processes during emergency management.

3. Translating forest fire behaviour into a Bayesian Network

The Bayesian model was developed in five steps:

1. preliminary general analysis of the problems and their implications;
2. break-down of the problems into several elementary models, which have been translated into elementary networks, to be connected later;
3. estimation of the quantitative relationships among the variables of each elementary network, starting from the availability of data and equations;
4. enhancement of the basic model into the final dynamic network made up of several time slices and its implementation into a VBA-based software tool;
5. validation of the final interconnected networks in a real scenario.

The basic Bayesian network developed for this application was intended as a snapshot, modelling what happens at a certain time in a pre-determined spatial cell. It has already been represented in Fig. 2, where square cells subdivide all the territory covered by the forest. The output variables estimated by the model and useful for forecasting where and how the fire will spread are the frontline speed, its direction and intensity. On the basis of these variables, it is possible to make an estimation of the moment in time when the cells will be completely burnt down and the fire will propagate towards one of the adjacent cells.

The input needed by the network may be divided into two categories, the first related to climatic variables:

- air temperature;
- wind intensity and direction;
- air humidity level;

and the second related to territorial features:

- forest type;
- forest wood maturity;
- ground slope;
- site or cell orientation.

It should be noticed that some of the input data may be collected directly by sensor measurements, because they are unsteady variables; while other data may be directly supplied by a GIS system.

3.1 Fire spread modelling

Fig. 4 shows the whole structure of one elementary network resembling the system logic underlying the whole phenomena and analysed thanks to the support of Rothermel's surface fire spread models.

Starting from the work by (Rothermel, 1972), the surface spread models that he developed were simplified by (Scott & Burgan, 2005): the new set was conceived so as to simulate only surface fire behaviour at the flame front, but not the residual combustion that takes place after the flame front has passed. These sets were drawn up starting from empirical observations, fire behaviour simulations over a range of midflame wind speeds and several moisture scenarios, hence several environmental conditions were taken into consideration. Dotted lines connecting solid filled circles (adjacent to some of the variables) in Fig. 4 mark those nodes to be merged at the final release of the whole network.

Among the fuel models proposed by the cited authors, eight have been used in the application developed in this Chapter (Fig. 5). Their naming follows a well-known rule: NB means non-burnable, GR means grass, GS stands for grass-shrub, SH for shrub, TU is timber-understory, TL for timber litter and SB is the abbreviation for slash-blowdown.

The following factors are listed for each forest type: the fuel loading available, extinction humidity values, relationships between midflame wind speed, rate of spread and flame length.

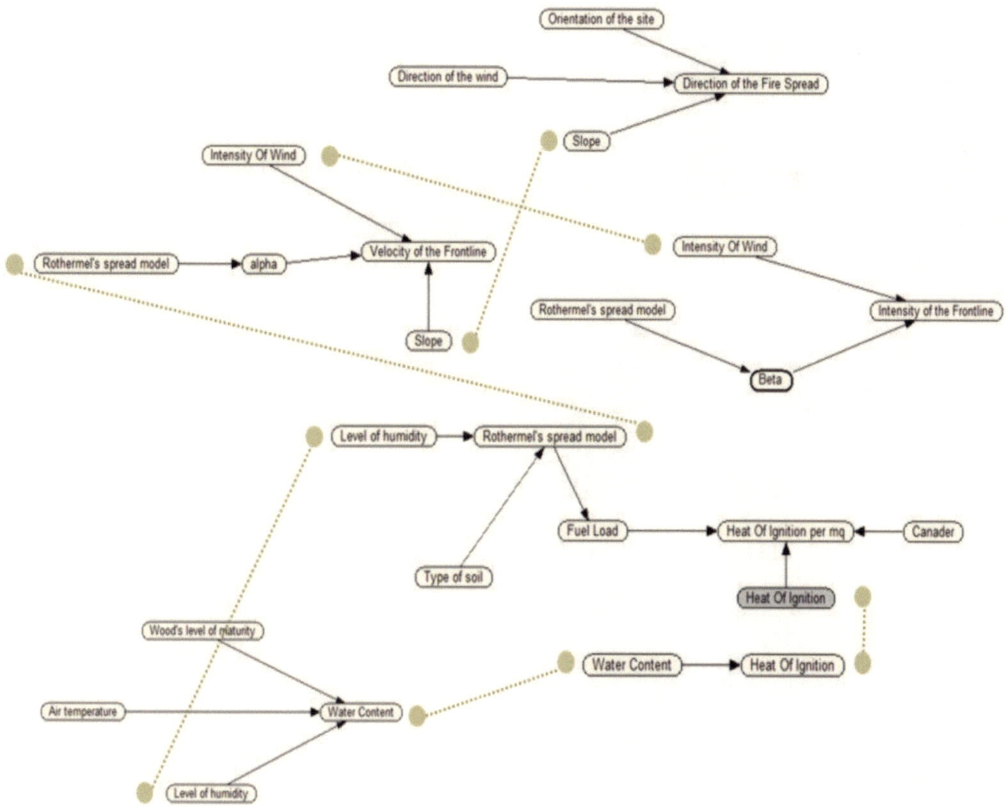

Fig. 4. Logic framework of the spatial cell of the Spatio-temporal Bayesian model

SH5: High Load, Dry Climate Shrub

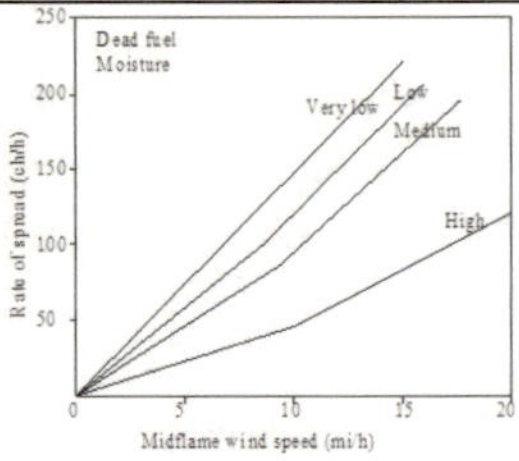

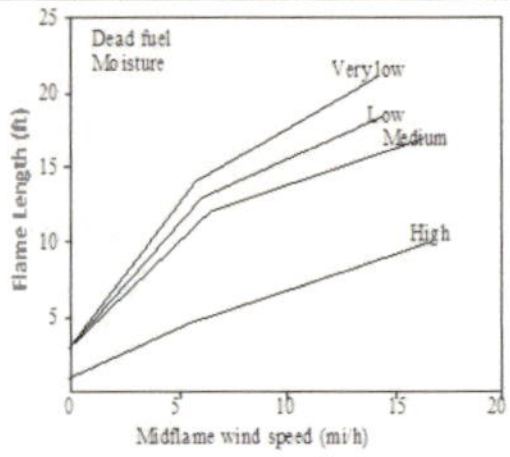

Fine fuel load(t/ac) 6.5 Extinction moisture content 15

SH8: High Load, Humid Climate Shrub

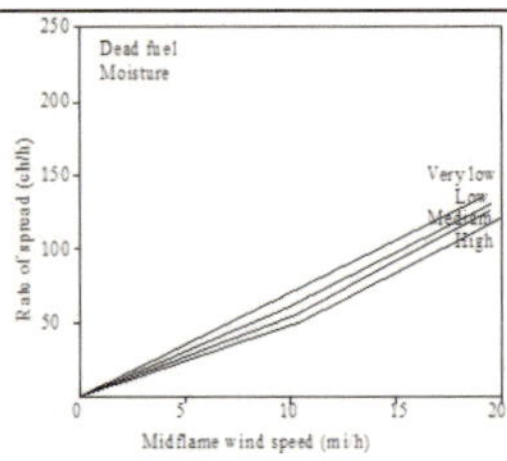

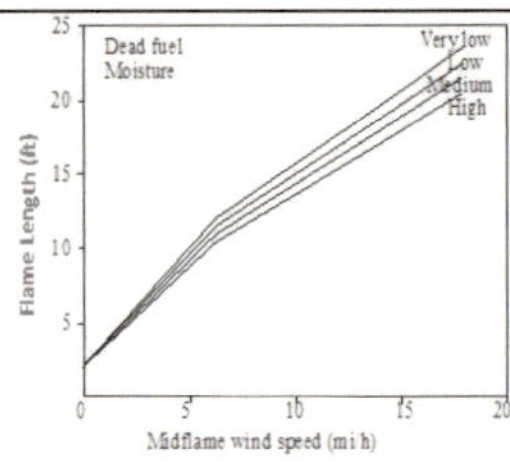

Fine fuel load(t/ac) 6.4 Extinction moisture content 40

TU2: Moderate Load, Humid Climate Timber-Shrub

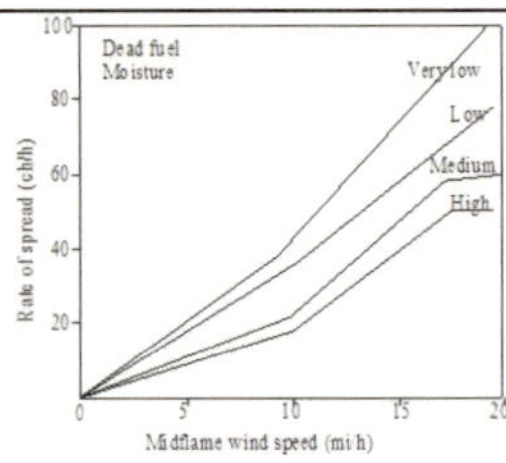

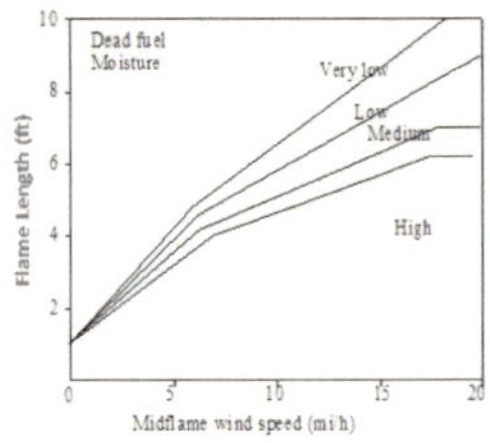

Fine fuel load(t/ac) 1.15 Extinction moisture content 30

TU4: Dwarf Conifer With Understory

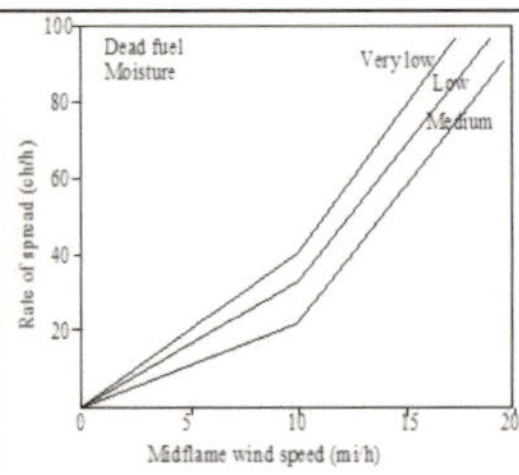

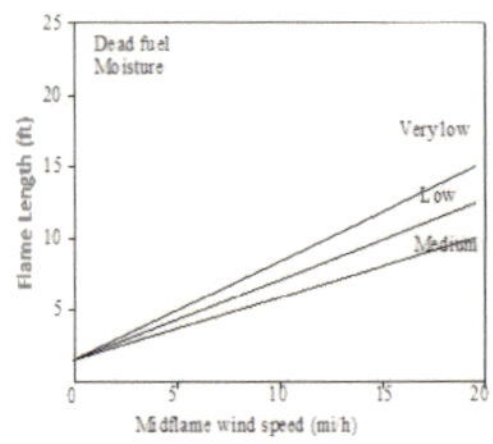

Fine fuel load(t/ac) 6.5 Extinction moisture content 12

TU5: Very High Load, Dry Climate Timber-Shrub

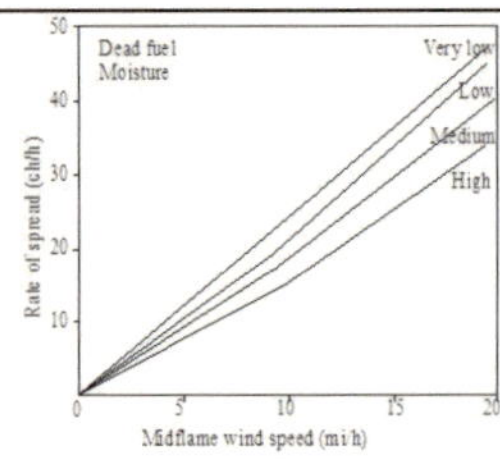

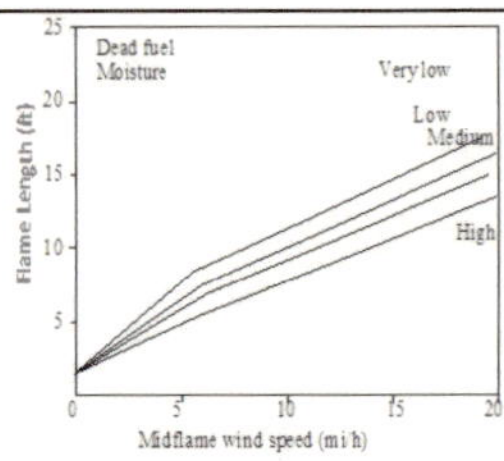

Fine fuel load(t/ac) 7.0 Extinction moisture content 25

TL6: Moderate Load Broadleaf Litter

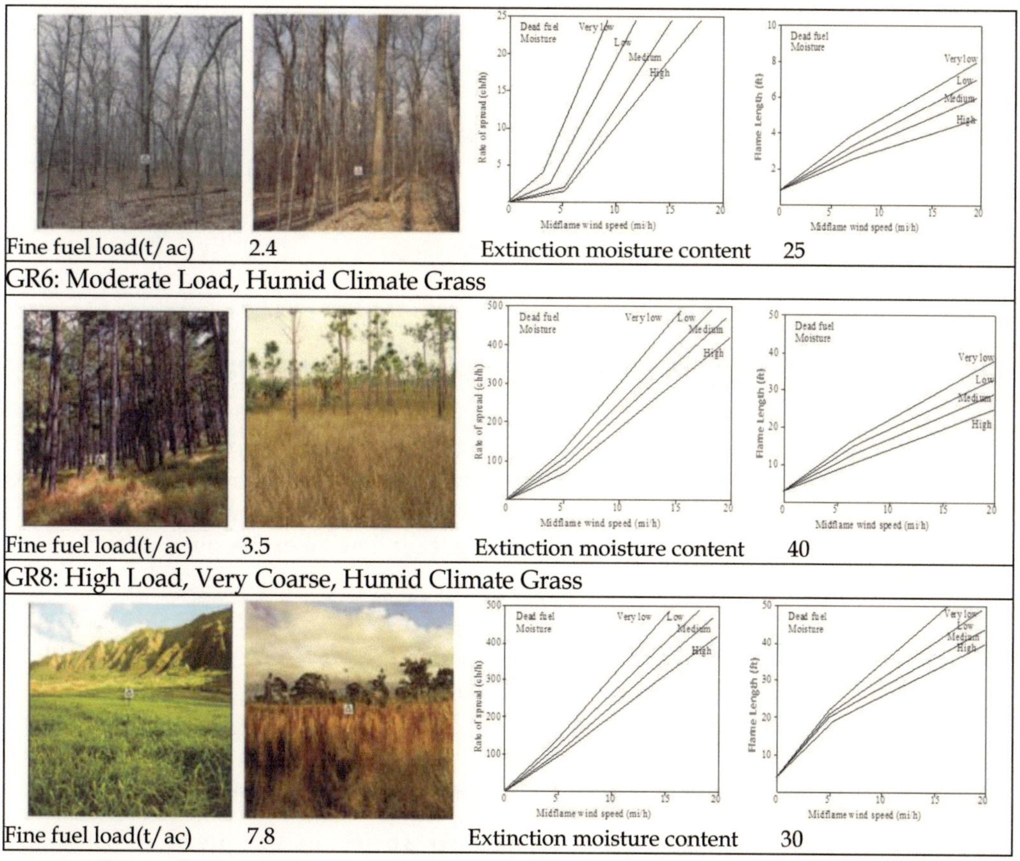

Fig. 5. Rothermel's models

Besides the aforedescribed models, other important analytical relations are those regarding the heat of ignition: this is the amount of heat which has to be released by the burning process so that the fire is not extinguished. It is strongly affected by the water content (hampering the process) and other parameters. However water content has been demonstrated to be the most important variable (it is tens or hundreds of times more important than other variables such as the initial temperature of the fuel), hence it is usually approximated as (Frandsen, 1972; Wilson, 1980):

$$Q_{ig} = 119.74 + 5.34 \cdot M_f \quad (2)$$

Where M_f is the percentage of water contained in the fuel loading, and is listed according to the type of ground. Such a variable directly affects the heat of ignition per unit of surface (measured in kJ/mq), which is the product:

$$Q_{ig,us} = Q_{ig} \cdot fuel\_loading \quad (3)$$

Fuel loading is dependent on Rothermel's surface fire spread models applicable to the type of ground, which are repeated for convenience: SH5, SH8, TU2, TU4, TU5, TL6, GR6, GR8. Hence the network will select all the right values, once the users have chosen the forest type under analysis.

Flame intensity is suggested by the same authors as the following expression:

$$I_f = 273 \cdot (2 + Wind\_intensity \cdot Beta)^{2.17} \quad (4)$$

where I_f is the flame length, which the models always report as dependent on wind intensity, "beta" is the slope of the linear model best approximating the average behaviour (see right diagrams in Fig. 5).

Frontline flame speed is given by:

$$S_f = Alfa \cdot Wind\_intensity + 12 \cdot (Slope/100)^2 \quad (5)$$

where "alfa" is the slope of the linear relationship best approximating fire behaviour (see left diagrams in Fig. 5).

Given the chance to perform active fighting against fire propagation, in this model the use of Canadairs has been considered and modelled with the data available in literature. It is supposed that water sprayed over fires increases the heat of ignition by the amount needed to completely evaporate the water, computed using the following expression:

$$Q_{ig} = 116 \cdot 1.055 \cdot M_r \quad (6)$$

where M_r is the amount of water falling on the ground. In order to give some practical tips, the Canadair model CL125 sprays a water strip as large as 20 x 85 = 1700 m^2 and ensures 3.20 l/m^2 on the ground.

3.2 Methodology for the development of the qualitative Bayesian model

The qualitative interpretation of the equations presented in the previous paragraph allowed us to build all the fragments of the network. More information about this will be given in 4.1.

Fig. 6. Bayesian network fragments relative to intensity and speed of the fire surface modelling

The fragments in Fig. 6 derive from equations 4 and 5.

Fig. 7-a pictures the fragment of the Bayesian network which estimates the fire surface spread direction. Empirical studies showed that this is related to wind direction, exposure and ground slope. Of course, when the slope and wind speed push the fire in opposite directions, spread is hampered.

The heat of ignition was derived directly from eq. (2), where its dependence on water content is clearly expressed (Fig. 7-b).

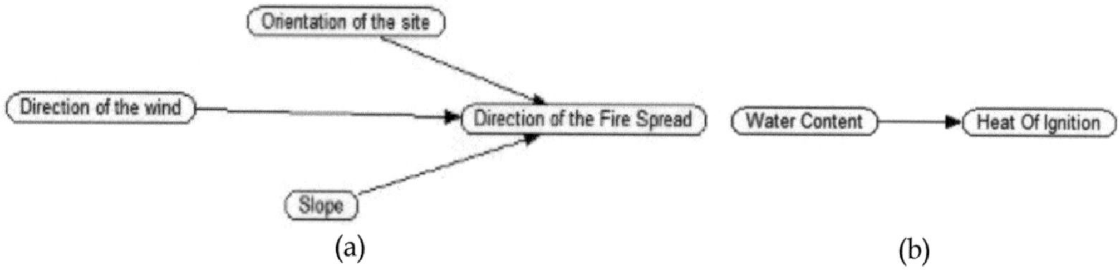

Fig. 7. Fragments relative to fire surface spread direction and heat of ignition

This last variable has repercussions on the heat of ignition per unit of surface, which is also dependent on fuel loading, as in eq. (3). By also adding Rothermel's fire surface spread models mentioned in 3.1 to this fragment, it is possible to obtain the fragment in Fig. 8:

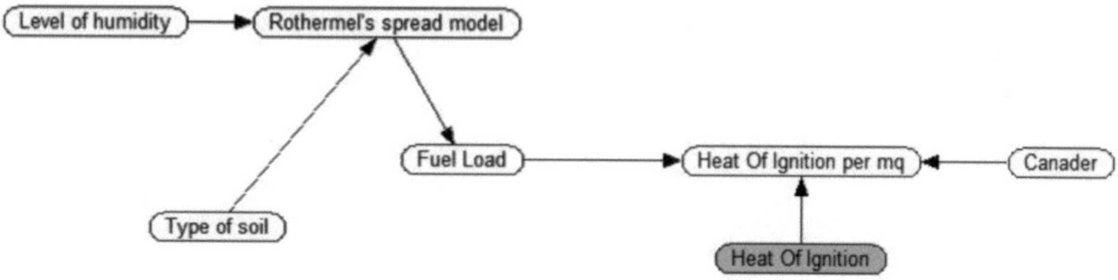

Fig. 8. Fragments relative to heat of ignition per unit of surface

This part of the model is also able to consider the possibility of active intervention through Canadairs, which would be useful to increase the heat of ignition and make the spread of fire less likely.

As a final integration to the proposed Bayesian networks, water content was considered dependent on other weather variables, which may be measured in real-time by sensors to speed up the process in emergency situations (Fig. 9).

At this stage, all the variables shared among different fragments were merged into the same variable to provide the whole final network properly instantiated (Fig. 4).

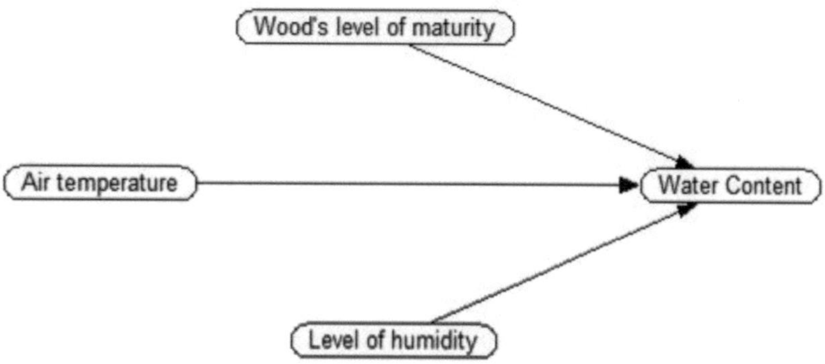

Fig. 9. Water content as a function of environmental variables

4. Model development

In this chapter we will first present the implementation of the quantitative relationships among the variables of a single cell of the whole Bayesian model. Spatial and temporal dynamics will then be introduced to produce the final model.

4.1 Implementation of quantitative relationships

The qualitative structure of the Bayesian model relative to a single spatial cell and described in paragraph 3 was integrated with the quantitative network through the use of several kinds of information. They may be roughly subdivided into two main groups:

- elementary cell networks whose CPTs have been derived directly from the observations of the available models and their translation into probability laws;
- analytical relationships explicitly available for all the networks, which have been translated into CPTs through the approach described in paragraph 2.1 (Hugin Expert, 2008).

In both cases a discretization process was carried out in an iterative way: starting from variables with a low number of states, their discretization was continuously refined until the data they provided were comparable to the numerical results obtained by running the Farsite™ software code, appropriately chosen for validation. All the elementary networks were developed in an Agena Risk™ environment.

In order to explain better, two examples are reported here. Figure 10 shows the fragment of cell Bayesian networks, regarding heat of ignition, dependent on water content according to equation (2) and resembling the qualitative fragment shown in figure 7-b.

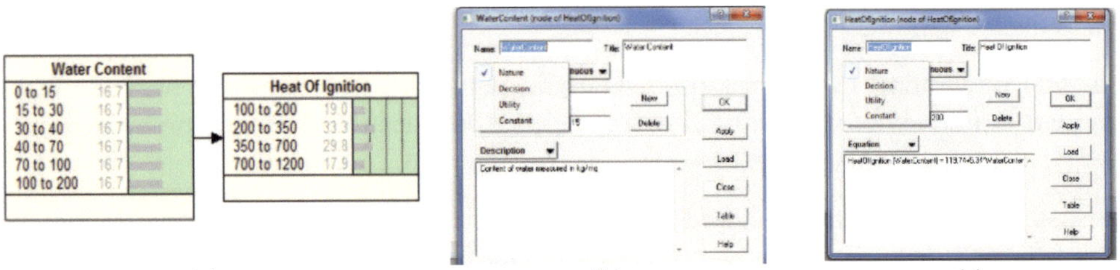

(a) (b) (c)

Fig. 10. Quantitative relationships among the variables

Once the discretization of the two variables has been accomplished and the analytical relationship has been inserted in the CPTs, the software computes the quantitative relationships as shown in the picture. Fig. 10-b and 10-c also show the discretization chosen for the variables and the analytical relationship inserted between the two in order to estimate the CPTs.

On the contrary, Fig. 11 represents a typical case in which conditional probabilities have been inserted directly from the reading of Rothermel's models. A direct reading of these data, as depicted in Fig. 11-a, is capable of estimating dependent variables starting from the reading of the independent ones. In this case once the type of soil is known (as in Fig. 8) the most fitting Rothermel's model can be selected and then the flame length or intensity or fuel load can be worked out. These data have been translated into CPTs of the network fragment in Fig. 11-b by selecting the state of the dependent variable following from the known state of the independent ones. Validation was carried out as in the previous case.

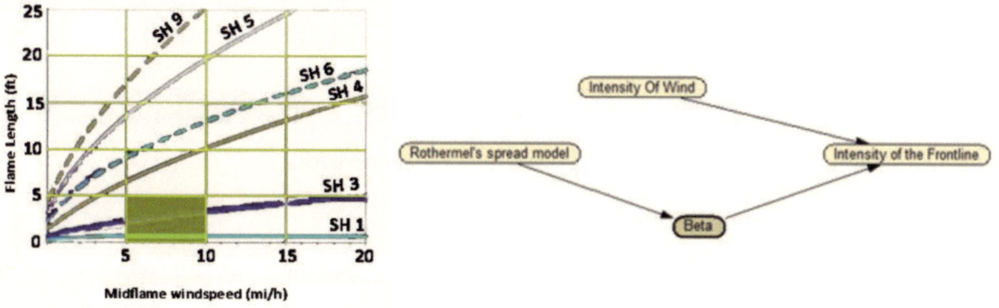

(a) (b)

Fig. 11. CPT evaluation using Rothermel's models

4.2 The control of spatial and temporal dynamics

The interpretation of a complex dynamic process, like fire propagation through woodlands, requires careful modelling of the spatio-temporal dynamics. In section 2.1 we have seen that the implementation of SDBN entails the discretization of both the spatial and the temporal variables. According to Niquist's theorem, the value of the discretization rates must be greater than twice the maximum frequency of variability of the phenomena in the spatial and in the temporal dimensions respectively. In the complex transition model shown in Fig. 2, this means that the time difference ΔT between two time slices must be less than half the propagation time of the fire inside a cell. This in turn depends, among other things, on the dimension of the cell itself, as we have seen in the previous section. Therefore in our case the spatial and the time sampling rates are strictly connected. External influencing factors, like the meteorological variables, must also be taken into account. Within each time slice the phenomena occurring are considered instantaneous. Therefore the external influencing factors are considered as constants in a single time slice, and changes can occur only between two different time slices. Hence the maximum frequency of variation in the external influencing phenomena must also be considered when defining the time discretization. As a general rule the time discretization must be the minimum between the fire propagation and the external meteorological dynamics.

The propagation time of fire inside a cell is worth further comment. As we have seen, the simulation of fire propogation must consider the time delay that occurs between the fire triggering in one cell and its propagation to the adjacent cells. This depends on many factors. For the sake of simplicity we assume that fire triggering heat is instantaneous. In this case the flame front reaches the next cells after a time delay that depends only on the cell dimensions and the speed of fire propagation. The speed of fire propagation depends, in turn, on the forest types which combine the speeds of both the low level grazing flames in the underwood and the high foliage flames. Therefore the use of forest type models avoided the implementation of layered networks, one for each fire type (please ref. to section 2.2), considerably simplifying the model. In conclusion, modelling the fire propagation temporal dynamics required the development of a specific subnetwork of the cell network that implements fire propagation tracking. This subnetwork is represented in Fig. 12 (the black node being inferred from the cell's network at each time slice as in Fig. 4).

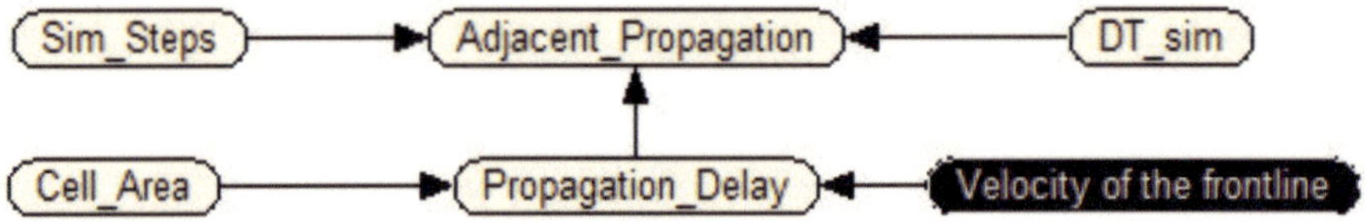

Fig. 12. Cell subnetwork implementing fire propagation control.

The subnetwork includes the following variables:

- *DT_sim:* is the simulation frequency, which is an evidence variable observed at the beginning of the simulation in seconds;
- *Sim_Steps*: is the number of simulation cycles, that is the number of time steps that have occurred from the beginning of the simulation step. The product of DT_sim and *Sim_Steps* gives the simulation time elapsed so far;
- *Propagation_Delay*: is the average time delay that occurs between the ignition of fire in the cells and its propagation to the adjacent cells;
- Adjacent_Propagation: is a Boolean variable that triggers when DT_sim ×Sim_Steps ≥ Propagation_Delay.

The propagation algorithm uses a stack of active cells, where it pushes in each simulation step the cells where the fire ignition process has started. We call these cells active. For each simulation cycle and for each active cell the *Sim-Steps* variable is observed (i.e. the simulation step is updated) and the *Adjacent_Propagation* variable is evaluated. If it is true then the adjacent cells are evaluated. For each adjacent cell, if the intensity of the flame front of the active cell is greater than the heat of ignition of the adjacent cell, and the adjacent cell has not been completely burnt down, then it is activated and pushed into the stack.

The implementation of this algorithm in the SDBN framework depicted in the previous section is straightforward, since the algorithm follows the transition rules mentioned in section 2.1. Consequently the simulation can be conducted step-by-step, instantiating the time transition frame in Fig. 12 at each step to the active cell under analysis by properly observing the evidence variables and by propagating the net. The same algorithm can be easily reversed by reversing the transition network links of the time frame in Fig. 12. In this way the fire propagation analysis can proceed backwards, in a diagnostic way. As will be shown in the next section this allows us to analyse the causes of the occurrence of fire in critical cells (e.g. small villages, roads, etc.) and to easily evaluate the effectiveness of action to contrast fire propagation, such as the use of Canadairs or fire-breaks in specific cells.

The complementarity of the SDBN-based simulation algorithm with standard simulation ones resides principally in the possibility to reverse the calculation. Backward fire propagation entails very well focused scenario analysis, since it goes from the effect to the causes. In fact, in order to obtain the same information that can easily be achieved with backward analysis many blind "generate and test" simulations are required with a forward algorithm. Backward analysis, even if affected by approximations, can give initial insight and guide the forward simulation by limiting the scope of the search space.

5. Model implementation

The complexity of the temporal model discussed in the previous sections, as well as its spatial extension to wide forest regions, usually leads to BN implementations containing thousands of nodes, resulting in unmanageable and impractical networks, that easily exceed the power of today's computers.

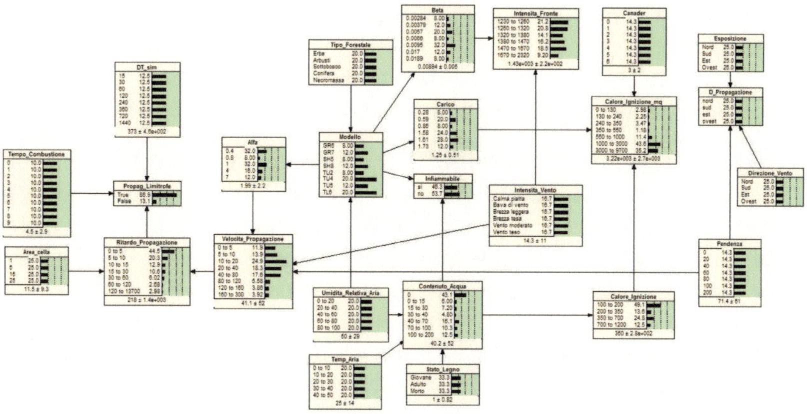

Fig. 13. Implementation of the single cell model with the Netica software

To solve such a complex issue we have used the property discussed in section 2, that allows for a stepwise implementation of the simulation algorithm. Fig. 13 shows the implementation of the single cell network in the Netica™ Software Programme, corresponding to the network discussed in chapter 4. The spatio-temporal transition network is shown in Fig. 14. This network contains 100 nodes and is able to represent large territories and support simulations for whatever time interval is necessary.

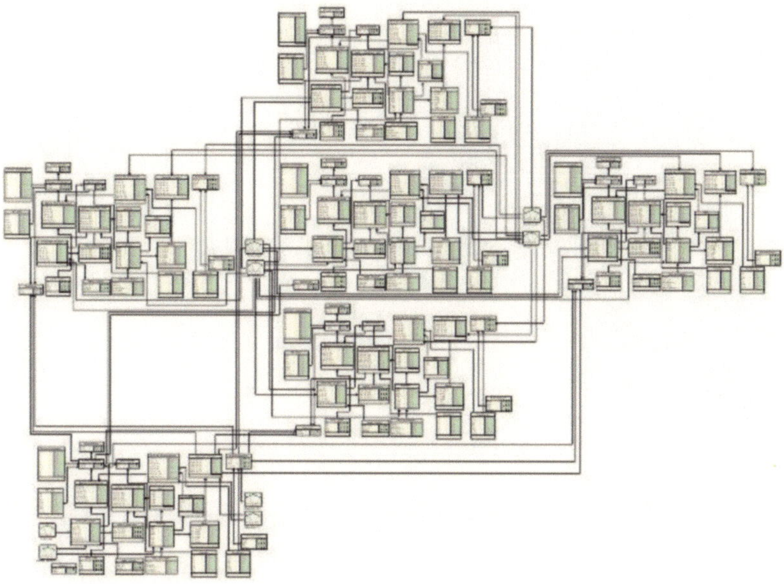

Fig. 14. Spatio-temporal transition network

5.1 The software interface

The simulator interface implemented in the prototype software is shown in Fig. 15. It consists of:

- a map of the territory which is divided into cells that can be selected with a click of the mouse;
- a cell window that allows for the input of the parameters of each cell (e.g. forest type, average slope, etc.);
- a simulator window that allows for the input of the simulation parameters (e.g. time step, time extension, type of simulation, wind direction and intensity, etc.).

It is worth noting the flag in this window that allows the selection between forward and backward simulation, as explained in the previous section. The step button makes the simulator proceed stepwise. In this way it is possible to change the meteorological conditions for each time slice. In this prototypal release of the software the grid that subdivides the map of the territory into cells is made up of a set of cells with a fixed square shape. This of course introduces some approximations in the forest type mapping, that, however, can be limited by using cells with smaller dimensions, resulting in a finer tessellation. This limitation will be overcome in future software releases.

5.2 Application to a real case study

The software prototype has been applied to analyse forest fire risk in the Esino-Frasassi forest district in the Marche region, in Italy, near Ancona. The forest area analysed extends for 3740 ha. It contains 6 main forest types. The area was subdivided into 154 cells of 25 ha each. For each cell the orographic and forestal parameters were inserted.

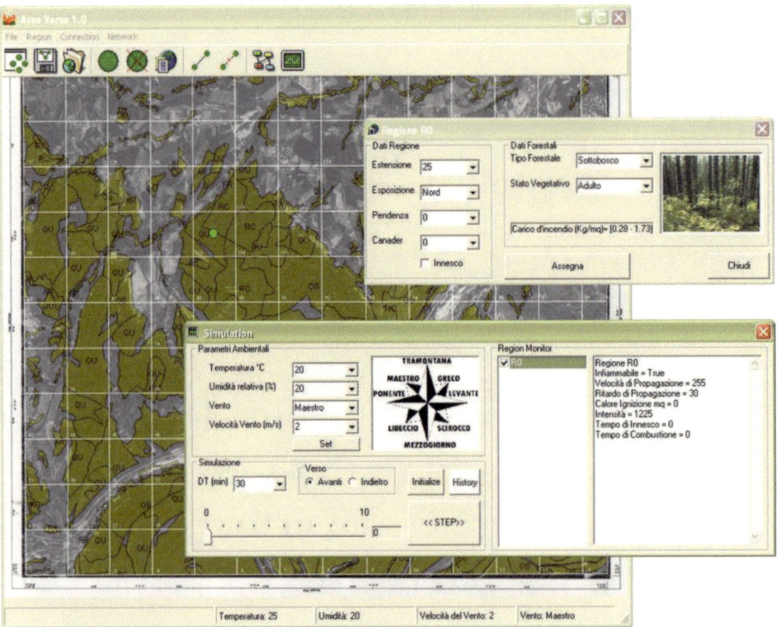

Fig. 15. The simple interface of the software prototype

Two kinds of simulations were carried out: forward standard simulations to evaluate both the risk of fire occurring when the fire breaks out in zones with a higher likelihood of ignition (roadsides, etc.) and the effectiveness of Canadair action; backward simulations to analyse the risk of disaster due to fire propagation involving areas with relatively high population density.

As an example Fig. 16 shows the dynamics of fire propagation in a forward simulation, with the fire breaking out in a west mountainside cell. The initial conditions in the triggering zone are: air temp 35°C, relative humidity equal to 40%, wind direction towards the south and wind speed 7m/s.

The direction and the extension of fire propagation depend on the forest type, on the meteorological conditions (quite severe in this case) and on the orography of the territory.

The software allows the effectiveness of the use of Canadairs and fire-breaks to be evaluated and, consequently, the optimization of their use. The use of Canadairs will increase the amount of heat necessary to light the fire in the cell, slowing down or even stopping the propagation of the fire. The use of Canadairs can be simulated by simply observing the usage rate in the Canadair node. Fig. 17 illustrates some steps in the same simulation shown in figure 16, but with the use of a Canadair.

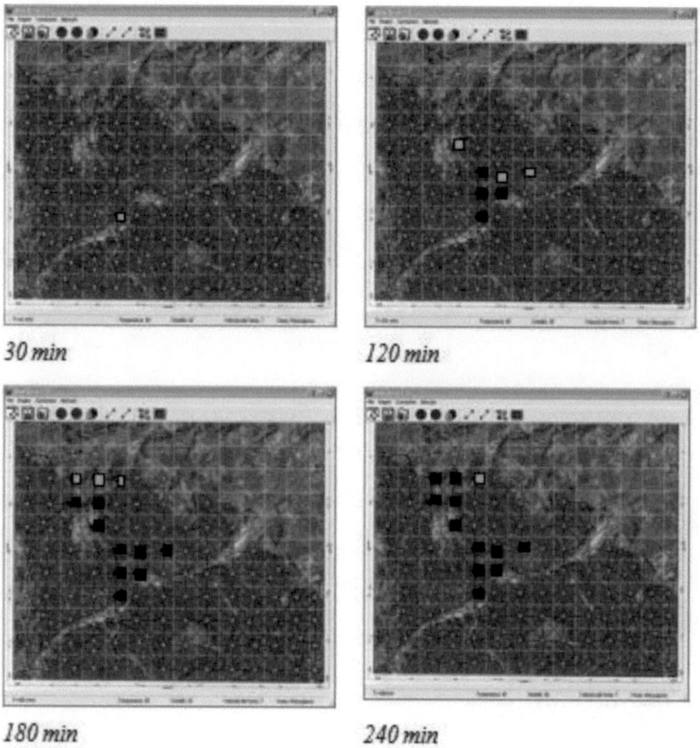

Fig. 16. Forward simulation of fire propagation. Gray = fire lighting phase, Black = final phase, the zone is almost burned down. The simulation time step is 30 minutes.

The example discussed so far has shown simulations that can be carried out, we should say even more accurately, with standard fire area simulation software. Having compared the results of the software prototype with the FARSITE™ simulator using the same Rothermel's models, *it* shows essentially the same dynamics, providing less accurate results in terms of space and time accuracy, due essentially to the discretization of the domains of the BN variables. Nevertheless the errors that have been introduced by the discretizations do not hinder the support to the scenario analyses that the system is aimed at.

The complementary role of BN-based fire area simulators, compared with standard simulation software packages, lies in their ability to proceed backwards. Proceeding backwards means that, once a key area is selected, it is possible to identify, as happens in a diagnostic process, all the possible paths that the fire can follow to reach the area, and to evaluate the effectiveness of risk mitigating action. Given the great number of possible paths once all the surrounding propagation cells are combined with all the possible meteorological conditions, the statistical analysis resulting from the adoption of the BN-based simulator seems to be the only feasible approach. Once the risk map for the area has been drawn (i.e. all the critical fire paths identified and the related risk mitigating policies defined), standard simulators can be applied to have more accurate evaluations of each critical path. Fig. 18 shows a backward simulation concerning the evaluation of fire risk for a cell which contains a small village. The initial conditions in the district are: air temp 35°C, relative humidity equal to 40%, wind direction towards the north and wind speed 2m/s.

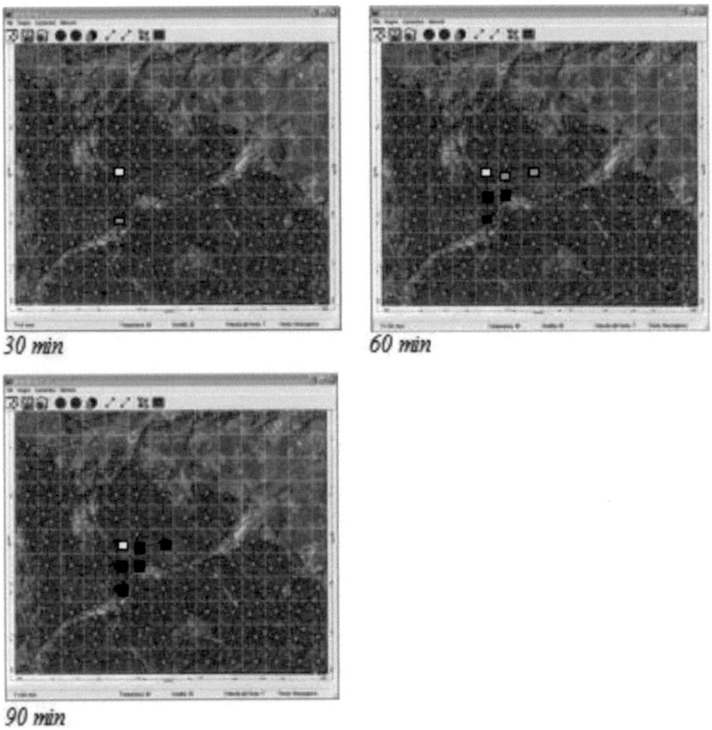

Fig. 17. Forward simulation of fire propagation with the use of a Canadair (white node). Gray = fire lighting phase, Black = final phase, the zone is almost burned down. The simulation time step is 30 minutes.

We can see that in these weather conditions the fire can reach the village essentially from one direction. Nevertheless this cell can be burned by all the adjacent cells, triggering a number of possible paths. Of course once the critical paths have been identified the simulation can be reversed and the use of Canadairs evaluated, as already illustrated.

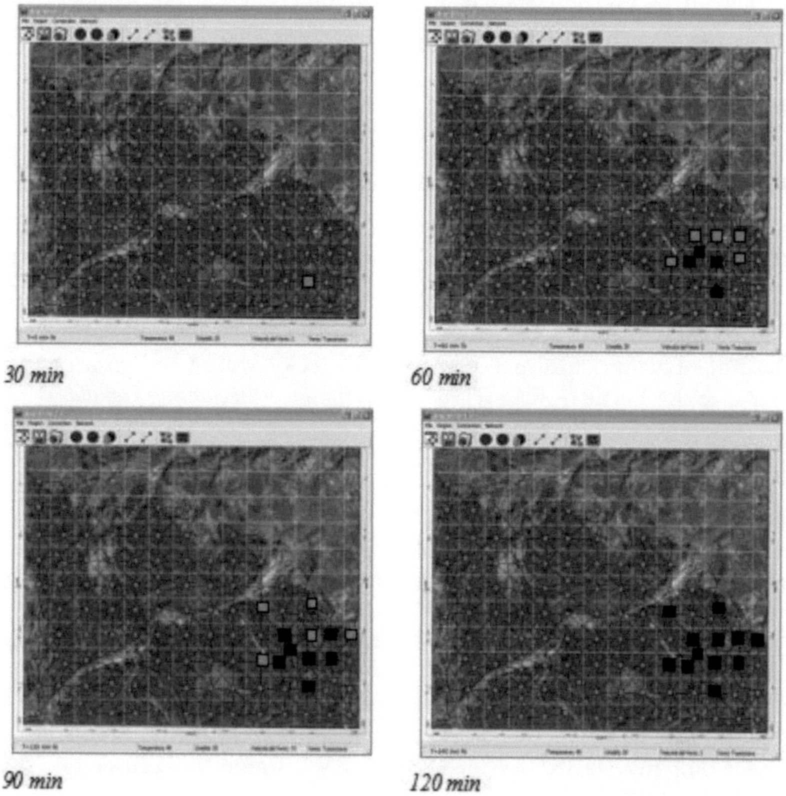

Fig. 18. Backward simulation of fire propagation and analysis of the risk concerning a cell which contains a small village. The location of the village is shown in the 30min simulation map.

6. Conclusions

The analysis provided in this paper was mainly devoted to showing the feasibility of an intelligent Decision Support System, capable of performing reliable analyses even in real-time during emergency situation management.

The particular case shown in this chapter concerns the development of a spatio-temporal Bayesian model, able to model both the spatial dynamics and the temporal evolution of forest fires. The former was tackled through the discretization of the forest into spatial cells, each corresponding to one elementary network. This discretization was used to create a spatial Bayesian network in the form of an Object-Oriented Network linking all the cells together. Time dependence was considered through the tool of Dynamic Bayesian Networks, which are able to quantify transition models between any one time slice and the next.

In order to make the models available for use by non-experts, everything was implemented into a VBA-based prototypal software tool, where a stepwise implementation of the simulation algorithm has been assumed. This means that any cell can influence only its adjacent ones, hence a limited number of cell networks must be considered at each simulation step.

The successful validation of the whole model has shown not only the reliability of the quantitative relationships implemented therein, but also the validity of assumptions regarding the stepwise spread of fire and spatial discretization. Bayesian Networks have also been proved capable of propagating complex evidence and managing a high number of variables.

7. Acknowledgment

The authors wish to warmly thank Eng. Diego Centanni for his operative involvement in the development of the Bayesian Networks described in this paper.

8. References

Andrews, P. L., Bevins, C. D., Seli, R. C. (2005). *BehavePlus fire modeling system, version 3.0 - User's Guide*, Gen. Tech. Rep. RMRS-GTR-106WWW Revised. Ogden, UT: Department of Agriculture, Forest Service, Rocky Mountain Research Station

Andrews, P.A. & Rothermel, R.C. (1982). *Charts for interpreting wildland fire behavior characteristics*, Gen. Tech. Rep. INT-131. Ogden, UT: U.S. Department of Agriculture, Forest Service, Intermountain Forest and Range Experiment Station

Brown A., Davis K. P. (1973). *Forest fire control and use* (2nd ed.), McGraw-Hill, ISBN: 0-908920-64-4, New York, NY

Hugin Expert (2008) *Hugin API – Reference Manual*, version 7.0, Hugin Expert A/S

Frandsen, W.H. (1972). *The effective heating of fuel particles ahead of a spreading fire*, USDA Forest service, Intermountain Forest and Range Exp. Sta., Ogden, Utah

Jensen, F. V. (1996). *An Introduction to Bayesian Networks*, UCL Press, ISBN: 1857283225, London

Kjaerulff, Uffe B., Madsen, Anders L. (2008) Bayesian Networks and Influence Diagrams: A Guide to Construction and Analysis, Springer, ISBN 978-0-387-74100-0

Korb, K. B., Nicholson, A. E. (2004). *Bayesian Artificial Intelligence*, Chapmann and Hall/CRC Press Company, ISBN: 1-58488-387-1, Boca Raton - London - New York - Washington DC

Luke, R.H. & McArthur, A.G. (1978). *Bushfires in Australia*, Australian Govt. Pub. Service, ISBN 0642023417, Canberra

Naticchia, B., Fernandez-Gonzalez, A., Carbonari A. (2007). Bayesian Network model for the design of roofpond equipped buildings. *International Journal of Energy and Buildings*, Vol. 39, No.3, pp. 258-272, ISSN: 0378-7788

Neapolitan, R.E. (2004). *Learning Bayesian Networks*, Prentice Hall, ISBN: 9780130125347, NJ

Pearl, J. (1988). *Probabilistic reasoning in intelligent systems : networks of plausible inferences* (2nd ed.), Morgan Kaufmann, ISBN: 1-55860-479-0, San Mateo, California

Rothermel, R.C. (1972). *A mathematical model for predicting fre spread in wildland fuels*, USDA Forest Service

Scott J.H., Burgan R.E. (2005). *Standard Fire Behavior Fuel Models: A Comprehensive Set for Use with Rothermel's Surface Fire Spread Model*, United States Department of Agriculture, Forest Service, Rocky Mountain Research Station

Tucker, A. & Liu, X. (2004). Learning dynamic Bayesian networks from multivariate time series with changing dependencies, *Proceedings of the Fifth International Symposium on Intelligent Data Analysis* (IDA 2003) *"Advances in Intelligent Data Analysis V"*, Lecture Notes in Computer Science 2810, 100–110, Springer

Wilson, R. (1980). Reformulation of Fire Spread Equations in SI units. *USDA Forest Service, Research Note INT-292*, Ogden, Utah

5

Probabilistic Inference for Hybrid Bayesian Networks

Wei Sun and Kuo-Chu Chang
Department of Systems Engineering & Operations Research
George Mason University, Fairfax, VA
USA

1. Introduction

A Bayesian network (BN) Charniak (1991) Pearl (1988) Jensen (1996) Neapolitan (1990) is a directed acyclic graph (DAG) consisting of nodes and arrows, in which node represents random variables, and arrow represents dependence relationship between connected nodes in the sense of the probabilistic, deterministic, or functional. Each node in BN has a specified conditional probability distribution (CPD), where all CPDs together parameterize the model. BNs have been used as powerful probabilistic knowledge models for decision support under uncertainty over a few decades, with numerous applications such as classification, medical diagnosis, bioinformatics, speech recognition, etc. One of the most important features BN has is the factorization of the joint probability space, so that conditional independence can be exploited to simplify modeling and save computations. However, BN model is only useful when combined with efficient algorithms for inference.

Over decades after the first Bayesian network (BN) was introduced in early 1980s, a number of inference algorithms have been reported in the literature. However, for hybrid Bayesian networks with both discrete and continuous variables, which are usually inevitable in modeling real-life problems, inference task has many difficulties and open issues. This chapter focuses on introducing the state-of-the-art hybrid inference methods in the literature. Particularly, we take scalability as a very important aspect and intend to provide the reader the opportunities to get the efficent inference methods under different circumstances.

The simplest hybrid Bayesian network is called Conditional Linear Gaussian (CLG) and it is a hybrid model for which exact inference can be performed by the Junction Tree (JT) algorithm Lauritzen (1992). However, JT and all of other exact inference algorithms have the complexity of being, in general, exponential to the size of the largest clique of the strongly triangulated graph. For a hybird BN model, there surely exists hybrid cliques that including all of discrete parent nodes for a connected continuous subgraph and at least one continuous node from the subgraph, which is usually the largest clique. Therefore, in most of real applications, exact inference is intractable.

For a general hybrid Bayesnet, due to the difficult issues such as the heterogeneity of variables, arbitrary densities involved, and possibly any functional relationships, with network topologies that may have discrete variables as parents of continuous nodes, we have to rely on approximate inference with tradeoff in accuracy against complexity. To this end, there are several main categories of approximate algorithms:

1. Generalized Junction Tree algorithm: when the complexity of a network is beyond the capability of traditional JT, Koller et al. (1999) proposed a general algorithm under the framework of Junction Tree, but using approximate clique potentials to do the clique tree propagation. It involves approximate inference algorithms to estimate the densities in each clique. Further, for hybrid network with the structure such that continuous variable has discrete children, Shenoy (2006) introduces a way to convert the model into a network with CLG structure, and then use Gaussian mixtures to approximate clique potentials for inference under Junction Tree framework. Another interesting method to approximate densities is to use truncated exponential Cobb & Shenoy (2006).
2. Hybrid loopy belief propagation: also know as message passing, the first belief propagation algorithm was proposed by Pearl in 1980s Pearl (1988), to provide exact inference for discrete polytree BN model. When there is any loop in the network, it becomes loopy belief propagation and provides accurate approximate solutions empirically for discrete networks Murphy et al. (1999). In hybrid case, Yuan & Druzdzel (2006) proposed a computationally extensive approach combining nonparametric belief propagation Sudderth et al. (2003), numerical integration, and density estimation techniques to pass messages between any types of variables.
3. Monte Carlo: importance sampling methods, such as Likelihood Weighting Fung & Chang (1989), Shachter & Peot (1999), are model-free algorithms, but usually have difficulty in dealing with unlikely evidence. The state-of-the-art importance sampling algorithms are AIS-BN Cheng & Druzdzel (2000) and EPIS-BN Yuan & Druzdzel (2007). Unfortunately, both work for discrete networks only. For hybrid BN models, any approximate results obtained by algorithms in the first two categories can be certainly used as the importance functions for efficient sampling process. Other sampling methods include Markov Chain Monte Carlo (MCMC) Gilks et al. (1996), Gamerman & Lopes (2006).
4. Variational methods: by formulating probabilistic inference into an optimization problem, variational methods provide another perspective for approximation solutions (Wainwright & Jordan (2008)).

We are particularly interested in the message passing framework because of its simplicity of implementation and good empirical performance, and more importantly, its distributed nature of inference. Without the computational burden of numerical integration, we proposed a partitioned message passing algorithm in Sun & Chang (2009), using interface nodes to separate the original network into sub-networks. Each sub-network contains only one type of variables, either discrete or continuous. We then conduct message passing separately within each sub-network. Finally, messages are fused together through interface nodes and the posterior distributions are computed based on final messages. The advantage of the partitioned message passing method is that it is easier to accommodate an efficient algorithm for inference within homogeneous sub-networks. On the other hand, a disadvantage is that we have to conduct inference conditioning on all the discrete parent nodes (i.e., interface nodes), for each connected continuous subgraph. Therefore, the algorithm has an exponential complexity proportional to the product of sizes of discrete parent nodes.

It is more desirable to have an unified message passing framework that allows direct message propagation between different types of variables for general hybrid Bayesian networks. We achieve this goal by deriving formulae for exchanging messages under all possible scenarios. Unscented transformation are used to tackle possible nonlinear functional relationships between continuous variables Julier (2002), Sun & Chang (2007a). For arbitrary densities, we proposed to use Gaussian mixture as the approximation before passing messages. The approach does not require any graph transformation, or any numerial integrations. In the framework, each node in the networks propagates messages to its neighbors. Messages are computed locally based on the node types under various circumstances without global knowledge. To maintain scalability, we also propose to use Gaussian mixture reduction techniques Kuo-Chu Chang & Smith (2010), Chang & Sun (2010), to limit the number of Gaussian components, while having the approximation error bounded each time. We term this new scalable and distributed approach Direct Message Passing for Hybrid Bayesian Network (DMP-HBN). Further, for general hybrid models with topology such that a discrete node may have continuous parents, one can always use Shenoy (2006) to convert the model into a network with CLG structure, then apply DMP-HBN for inference. This algorithm is able to provide an exact solution for polytree CLGs, and approximate solution by loopy propagation for general hybrid models.

In the rest of this chapter, we focus on describe the details of DMP-HBN. At the end of this chatper, we will also briefly discuss an up-to-date method to find the most probable explanations (MPE) for hybrid Bayesian networks.

2. Direct message passing

This section describes DMP-HBN algorithm in detail. We first briefly review Pearl's original message passing algorithm. We then extend it for general hybrid models.

2.1 Pearl's message passing algorithm

Recall that in a polytree network, any node X d–separates evidence into $\{\mathbf{e}^+, \mathbf{e}^-\}$, where $\mathbf{e}^+$ and $\mathbf{e}^-$ are evidence from the sub-network "above" X and "below" X respectively. Every node in the network maintains two values called λ and π. The λ value of X is the likelihood, defined as:

$$\lambda(X) = P(\mathbf{e}_X^-|X) \tag{1}$$

The π value of X, defined as:

$$\pi(X) = P(X|\, \mathbf{e}_X^+) \tag{2}$$

is the conditional probability distribution of X given $\mathbf{e}_X^+$. It is easy to see that the belief of a node X given all evidence is just the normalized product of its λ and π values:

$$\begin{aligned}
BEL(X) &= P(X|\mathbf{e}) = P(X|\mathbf{e}_X^+, \mathbf{e}_X^-) \\
&= \frac{P(\mathbf{e}_X^-|X, \mathbf{e}_X^+)P(X|\mathbf{e}_X^+)P(\mathbf{e}_X^+)}{P(\mathbf{e}_X^+, \mathbf{e}_X^-)} \\
&= \alpha P(\mathbf{e}_X^-|X)P(X|\mathbf{e}_X^+) \\
&= \alpha\lambda(X)\pi(X)
\end{aligned} \tag{3}$$

where α is a normalizing constant. In message passing, every node sends λ messages to each of its parents and π messages to each of its children. Based on its received messages, every node updates its λ and π values correspondingly. The general message propagation equations of Pearl's algorithm are the following Pearl (1988):

$$\pi(X) = \sum_{\mathbf{T}} P(X|\,\mathbf{T}) \prod_{i=1}^{m} \pi_X(T_i) \tag{4}$$

$$\lambda(X) = \prod_{j=1}^{n} \lambda_{Y_j}(X) \tag{5}$$

$$\pi_{Y_j}(X) = \alpha \left[\prod_{k \neq j} \lambda_{Y_k}(X) \right] \pi(X) \tag{6}$$

$$\lambda_X(T_i) = \sum_{X} \lambda(X) \sum_{T_k:\, k \neq i} P(X|\mathbf{T}) \prod_{k \neq i} \pi_X(T_k) \tag{7}$$

where $T = (T_1, T_2, ..., T_n)$ are the parents of node X; $Y = (Y_1, Y_2, ..., Y_m)$ are children of node X; $\lambda_{Y_j}(X)$ is the λ message node X receives from its child Y_j, $\lambda_X(T_i)$ is the λ message X sends to its parent T_i; $\pi_X(T_i)$ is the π message node X receives from its parent T_i, $\pi_{Y_j}(X)$ is the π message X sends to its child Y_j; and α is a normalizing constant.

Equations (4) to (7) are recursive equations, so we need to initialize messages properly to start the message propagation. Again, Pearl's algorithm is originally designed for discrete polytree networks, so these propagation equations are for computing discrete probabilities. When Pearl's algorithm is applied to a pure discrete polytree network, the messages propagated are exact and so are the beliefs of all nodes after receiving all messages. For pure continuous networks with arbitrary distributions, we proposed a method called Unscented Message Passing Sun & Chang (2007a) using a similar framework with different message representations and a new corresponding computation method. However, with both discrete and continuous variables in the model, passing messages directly between different types of variables requires additional techniques.

2.2 Direct message passing between discrete and continuous variables

We focus our research in this paper to the type of hybrid Bayesian networks that have the same network structure as the CLG, named conditional hybrid model (CHM). In a CHM, a continuous node is not allowed to have any discrete child, while it may have arbitrary distributions and nonlinear relationships between variables. We believe that it is not difficult to extend our algorithm to general hybrid models with arbitrary network structure. Therefore in a CHM, the only case we need to consider when exchanging message between different types of variables is when a continuous node has discrete parents. Without loss of generality, suppose that we have a typical hybrid CPD involving a continuous node X with a discrete parent node D and a continuous parent node U, as shown in 1. Messages sent between these nodes are: (1) π message from D to X, denoted as $\pi_X(D)$; (2) π message from U to X, denoted as $\pi_X(U)$; (3) λ message from X to D, denoted as $\lambda_X(D)$; and (4) λ message from X to U, denoted as $\lambda_X(U)$. In addition, each node needs to maintain its λ and π values.

Let us look at these messages one by one, and derive their corresponding formula based on Pearl's traditional message passing mechanism. First, recall from Equation (6), $\pi_X(D)$ can be computed by substitution:

$$\pi_X(D) = \alpha \left[\prod_{child \neq X} \lambda_{child}(D) \right] \pi(D) \tag{8}$$

where $\lambda_{child}(D)$ is λ message sent to D from each of its children except X, and $\pi(D)$ is the easily computed message sent from the discrete sub-network "above" D. Note that $\lambda_{child}(D)$ is always in the form of a discrete vector. After normalizing, $\pi_X(D)$ is a discrete probability distribution serving as the mixing prior for a Gaussian mixture.

Similarly, but in a different form, $\pi_X(U)$ can be computed as:

$$\pi_X(U) = \alpha \left[\prod_{child \neq X} \lambda_{child}(U) \right] \pi(U) \tag{9}$$

where $\lambda_{child}(U)$ are λ messages sent to U from its continuous children other than X. These λ messages are continuous messages in the form of Gaussian mixtures. $\pi(U)$ is π value of U, and its computation depends on the type of parent nodes it has. The generalized computation of $\pi(X)$ will be described in the next paragraph. Finally, the resulting $\pi_X(U)$ is a normalized product of Gaussian mixtures, resulting in another Gaussian mixture with a greater number of components.

Now for $\pi(X)$, by applying Equation (4) with integral replacing summation for continuous variable, we have,

$$\begin{aligned} \pi(X) &= \sum_D \int_U P(X|D,U)\pi_X(D)\pi_X(U)dU \\ &= \sum_D \left[\pi_X(D) \int_U P(X|D,U)\pi_X(U)dU \right] \end{aligned} \tag{10}$$

where $\pi_X(D)$ and $\pi_X(U)$ are π messages sent from D and U respectively. For a given $D = d$, $P(X|D = d, U)$ defines a probabilistic functional relationship between X and its continuous parent U. The integral of $P(X|D = d, U)\pi_X(U)$ over U is equivalent to a functional transformation of $\pi_X(U)$, which is a continuous message in the form of a Gaussian mixture. In this functional transformation process, we pass each Gaussian component individually to

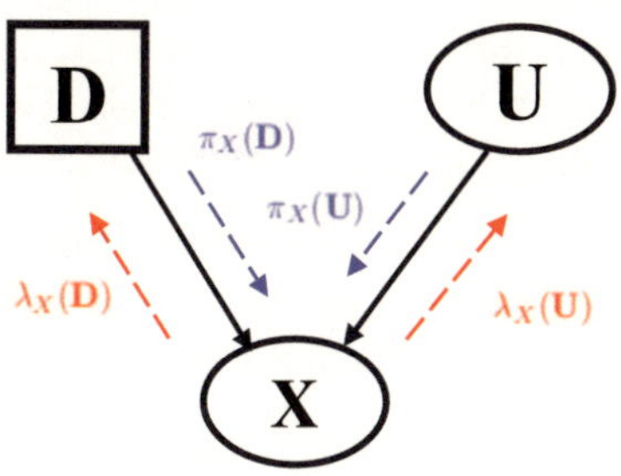

Fig. 1. A typical node with hybrid CPD — continuous node X has discrete parent D and continuous parent U.

form a new Gaussian mixture. Essentially, $\pi(X)$ is a mixture of continuous distributions weighted by $\pi_X(D)$. To avoid the potential for growing complexity of the message, it is possible to approximate the mixture with a single Gaussian density or a Gaussian mixture with fewer components.

$\lambda(X)$ is relatively straightforward to compute as it is the product of λ messages from each of its children, which must be continuous variables due to the network structure restriction. However, since we represent a continuous message as a Gaussian mixture, the product of a set of Gaussian mixtures will be another Gaussian mixture with increased number of components.

Let us now turn to the computation of messages sent from X to its parents D and U. As shown in Equation (7), λ message sent to its parents is essentially an inverse functional transformation of the product of the λ value of the node itself and the π messages sent from all of its other parents via the function defined in the CPD of X. It can be derived as,

$$\lambda_X(D=d) = \int_X \lambda(X) \int_U P(X|D=d,U)\pi_X(U)dUdX \tag{11}$$

where $\int_U P(X|D = d,U)\pi_X(U)dU$ is a functional transformation of a distribution over U into a distribution over X. Further, multiplying by $\lambda(X)$ and integrating over X, results in a non-negative constant, serving as a likelihood of X given $D = d$.

Similarly, the λ message sent from X to its continuous parent U can be expressed as:

$$\begin{aligned}\lambda_X(U) &= \int_X \lambda(X) \sum_D P(X|D,U)\pi_X(D)dX \\ &= \sum_D \left[\pi_X(D) \int_X \lambda(X)P(X|D,U)dX\right]\end{aligned} \tag{12}$$

Note that $\int_X \lambda(X)P(X|D,U)dX$ is an integral of the product of X's λ value and its conditional probability distribution; this integral is over X itself. Therefore it results in a density estimate of its parent multiplied by a coefficient. This coefficient is very critical in computing mixing priors with $\pi_X(D)$ when there is more than one component in the mixture distributions.

Equations (8) to (12) form a baseline for computing messages between discrete and continuous variables. Along with the well-defined formulae for computing messages between the same types of variables, they together provide an unified message passing framework for hybrid Bayesian network models. When the network is a polytree, messages propagated between nodes are exact and so the beliefs. When there are loops in the network, DMP-BN still works in the same way as so-called loopy propagation but provides approximate solution.

To illustrate the algorithm, next we describe in detail the computing process of message passing with a concrete 5-node polytree CLG called *Poly5CLG*. The network structure of *Poly5CLG* is shown in Figure 2. It consists of 2 discrete node T, C and 3 continuous nodes Y, W, Z. We assume binary discrete nodes and scalar Gaussian continuous nodes in the model. The corresponding CPDs are specified in Figure 3. Suppose leaf nodes C, Z are observable evidence and they are instaniated as state 1, and 5.5 respectively.

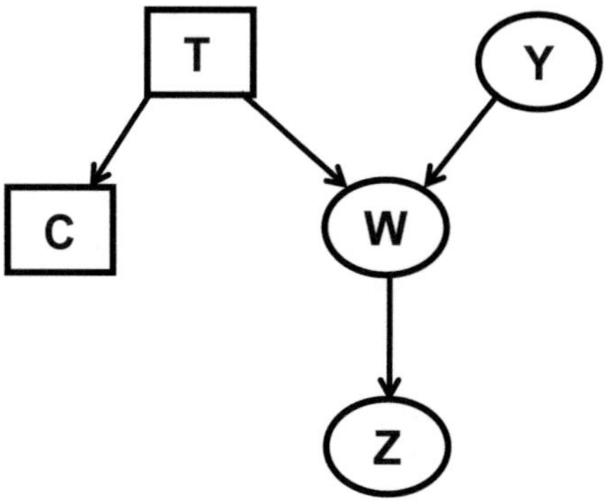

Fig. 2. *Poly5CLG*: A demo CLG model consisting of 2 discrete nodes T, C and 3 continuous nodes Y, W, Z.

T: $Pr(T=1)=0.5,\ Pr(T=2)=0.5$

C:

T	$T=1$	$T=2$
$Pr(C=1\|T)$	0.8	0.3
$Pr(C=2\|T)$	0.2	0.7

Y: $p(Y)=\mathcal{N}(10,1)$

W: $p(W|T=1)=\mathcal{N}(-1+Y,1)$, $p(W|T=2)=\mathcal{N}(1+Y,1)$

Z: $p(Z)=\mathcal{N}(0.5W,1)$

Fig. 3. Nodes CPDs for model *Poly5CLG*.

The algorithm is based on iterative computations. First, every node initializes its own λ, π values and messages propagated to its parents (λ messages) and children (π messages). Then, in each iteration, every node updates their λ, π values and messages, utill all of nodes converge to their steady beliefs. For ease of exposition, here we describe the computing process starting from evidence nodes towards their neighbors, and then propagating to other hidden nodes. Further since we only need to know the posterior distributions for hidden variables, we do not compute the messages back to evidence nodes. Starting with the messages from nodes C, Z, it is easily understood that,

$$\lambda(C) = \begin{bmatrix} 1 \\ 0 \end{bmatrix},\ \lambda(Z) = \begin{cases} \mu = 5.5 \\ \sigma^2 = 0 \end{cases},$$

where μ, σ^2 are mean and variance representing the continuous message. Then, the λ message sending from C to its parent T can be obtained as

$$\lambda_C(T) = \sum_C \lambda(C)P(C|T) = \sum_C \begin{bmatrix} 1\ 1 \\ 0\ 0 \end{bmatrix} \cdot \begin{bmatrix} 0.8\ 0.3 \\ 0.2\ 0.7 \end{bmatrix} = [0.8\ 0.3],$$

where . is the elementwise multiplication of matrices. Please note that computing λ message sending from Z to its parent W is more complicated and subtle:

$$\lambda_Z(W) = \int_Z \lambda(Z)P(Z|W)dZ.$$

In general, this is essentially an inverse functional transformation for estimating the original dependent variable based on the information of independent variable. Generally, let us assume that the CPD of $P(Z|W)$ is

$$P(Z|W) = \mathcal{N}(f(W), \sigma_0^2),$$

where $f(W)$ is an arbitrary deterministic function specifying the functional relationship between Z and W. Suppose that we now know Z is distributed as $\mathcal{N}(\mu_z, \sigma_z^2)$ (serving as $\lambda(Z)$). Then, $\lambda_Z(W)$ is actually the estimate of W based on this known information about Z, computed as:

$$\begin{aligned}
\lambda_Z(W) &= \int_Z \lambda(Z)P(Z|W)dZ \\
&= \int_Z \mathcal{N}(\mu_z, \sigma_z^2)\mathcal{N}(f(W), \sigma_0^2)dZ \\
&= \int_Z \frac{1}{\sqrt{2\pi}\sigma_z} \exp\left\{-\frac{(Z-\mu_z)^2}{2\sigma_z^2}\right\}. \\
&\quad \frac{1}{\sqrt{2\pi}\sigma_0} \exp\left\{-\frac{(Z-f(W))^2}{2\sigma_0^2}\right\} dZ \\
&= \frac{1}{2\pi\sigma_0\sigma_z} \int_Z \exp \\
&\quad \left\{-\frac{\sigma_0^2(Z-\mu_z)^2 + \sigma_z^2(Z-f(W))^2}{2\sigma_0^2\sigma_z^2}\right\} dZ
\end{aligned} \tag{13}$$

Let us denote the part of exponent in Equation (13) as **E**,

$$\mathbf{E} = \frac{\sigma_0^2(Z-\mu_z)^2 + \sigma_z^2(Z-f(W))^2}{2\sigma_0^2\sigma_z^2}.$$

E can be rearranged to be,

$$\begin{aligned}
\mathbf{E} &= \frac{Z^2 - \frac{2\sigma_0^2\mu_z + 2\sigma_z^2 f(W)}{\sigma_0^2+\sigma_z^2} Z + \frac{\sigma_0^2\mu_z^2 + \sigma_z^2 f^2(W)}{\sigma_0^2+\sigma_z^2}}{\frac{2\sigma_0^2\sigma_z^2}{\sigma_0^2+\sigma_z^2}} \\
&= \frac{(Z-\mathbf{U})^2}{\frac{2\sigma_0^2\sigma_z^2}{\sigma_0^2+\sigma_z^2}} + \frac{(\mu_z - f(W))^2}{2(\sigma_0^2+\sigma_z^2)},
\end{aligned} \tag{14}$$

where $\mathbf{U} = \frac{\sigma_0^2 \mu_z + \sigma_z^2 f(W)}{\sigma_0^2 + \sigma_z^2}$ is a constant relative to variable Z. Substituting (14) back into (13),

$$\begin{aligned}
\lambda_Z(W) &= \frac{1}{2\pi\sigma_0\sigma_z} \int_Z \exp\left\{ -\frac{(Z-\mathbf{U})^2}{\frac{2\sigma_0^2\sigma_z^2}{\sigma_0^2+\sigma_z^2}} \right\} \\
&\quad \exp\left\{ \frac{(\mu_z - f(W))^2}{2(\sigma_0^2+\sigma_z^2)} \right\} dZ \\
&= \frac{\sqrt{\sigma_0^2+\sigma_z^2}}{\sqrt{2\pi}\sigma_0\sigma_z} \int_Z \exp\left\{ -\frac{(Z-\mathbf{U})^2}{\frac{2\sigma_0^2\sigma_z^2}{\sigma_0^2+\sigma_z^2}} \right\} dZ \\
&\quad \frac{\sqrt{2\pi}\sigma_0\sigma_z}{\sqrt{\sigma_0^2+\sigma_z^2}} \frac{1}{2\pi\sigma_0\sigma_z} \exp\left\{ -\frac{(f(W)-\mu_z)^2}{2(\sigma_0^2+\sigma_z^2)} \right\} \\
&= \frac{1}{\sqrt{2\pi(\sigma_0^2+\sigma_z^2)}} \exp\left\{ -\frac{(f(W)-\mu_z)^2}{2(\sigma_0^2+\sigma_z^2)} \right\} \qquad (15) \\
&= \frac{1}{\sqrt{2\pi(\sigma_0^2+\sigma_z^2)}} \sqrt{2\pi}\sigma_w \frac{1}{\sqrt{2\pi}\sigma_w} \\
&\quad \exp\left\{ -\frac{(W-\mu_w)^2}{2\sigma_w^2} \right\} \\
&= \frac{\sigma_w}{\sqrt{\sigma_0^2+\sigma_z^2}} \mathcal{N}(\mu_w, \sigma_w^2), \qquad (16)
\end{aligned}$$

where μ_w, σ_w^2 are the mean and variance estimates for variable W, which always can be obtained by rearranging the exponent in (15) and they must be functions of $\mu_z, \sigma_0, \sigma_z$. In our algorithm, we use unscented transformation to estimate the post distributions for variables undergone nonlinear functions. Note that the constant coefficient $\frac{\sigma_w}{\sqrt{\sigma_0^2+\sigma_z^2}}$ must be part of λ message. It is very critical to keep the coefficient in place while the λ message is in the form of mixture distributions so that it can be updated with correct weights of the components. From (16), it also shows that the λ message is not a distribution. Instead, it is a probabilistic likelihood function. In a special case such that Z is observed as the value z ($\mu_z = z$ and $\sigma_z^2 = 0$), then Equation (15) can be simplified to:

$$\lambda_Z(W) = \frac{1}{\sqrt{2\pi\sigma_0^2}} \exp\left\{ -\frac{(f(W)-z)^2}{2\sigma_0^2} \right\} = \frac{\sigma_w}{\sigma_0} \mathcal{N}(\mu_w, \sigma_w^2), \qquad (17)$$

where μ_w is a function of z, and σ_w is a function of σ_0. It is straightforward to extend Equation (15), (16), and (17) for continuous node with multiple parents, by adding Gaussian terms from the continuous parents and functions given discrete parents.

Back to the concrete example, substituting actual functions and values into (17),

$$\begin{aligned}\lambda_Z(W) &= \frac{1}{\sqrt{2\pi} \times 1} \exp\left\{-\frac{(0.5W - 5.5)^2}{2 \times 1}\right\} \\ &= \frac{1}{\sqrt{2\pi}} \exp\left\{-\frac{(W - 11)^2}{2 \times 4}\right\} \\ &= 2\mathcal{N}(\mu_w = 11, \sigma_w^2 = 4). \end{aligned} \tag{18}$$

Since Z is the only child of W, from Equation (5), we have,

$$\lambda(W) = \lambda_Z(W) = 2\mathcal{N}(\mu_w = 11, \sigma_w^2 = 4).$$

For hidden root nodes T, Y, their π values are just their prior distributions,

$$\pi(T) = [0.5\ 0.5], \qquad \pi(Y) = \mathcal{N}(\mu_y = 10, \sigma_y^2 = 1).$$

Now we can compute π messages sending from T, Y to W respectively, according to Equation (8) and (9),

$$\pi_W(T) = \alpha\lambda_C(T)\pi(T) = \alpha[0.8\ 0.3].[0.5\ 0.5] = [0.7273\ 0.2727];$$
$$\pi_W(Y) = \pi(Y) = \mathcal{N}(\mu_y = 10, \sigma_y^2 = 1).$$

Then,

$$\begin{aligned}\pi(W) &= \sum_T \left[\pi_W(T) \int_Y P(W|T,Y)\pi_W(Y)dY\right] \\ &= 0.7273\mathcal{N}(\mu_w = 9, \sigma_w^2 = 2) + 0.2727\mathcal{N}(\mu_w = 11, \sigma_w^2 = 2),\end{aligned}$$

which is a Gaussian mixture. So far, W has received all of messages from its parents and children, so,

$$\begin{aligned}BEL(W) &= \alpha\ \lambda(W)\pi(W) \\ &= \alpha\ 2\mathcal{N}(11,4)\ [0.7273\mathcal{N}(9,2) + 0.2727\mathcal{N}(11,2)] \\ &= \alpha\ [2 * 0.7273 * 0.1167\mathcal{N}(9.6667, 1.3333)] + \\ &\qquad \alpha\ [2 * 0.2727 * 0.1629\mathcal{N}(11, 1.3333)] \\ &= 0.6564\mathcal{N}(9.6667, 1.3333) + 0.3436\mathcal{N}(11, 1.3333)\ .\end{aligned}$$

With (11) and (12), the λ messages sending from W to its parents are,

$$\begin{aligned}\lambda_W(T=1) &= \int_W \lambda(W) \int_Y P(W|T=1,Y)\pi_W(Y)dYdW \\ &= \int_W 2\mathcal{N}(11,4)\mathcal{N}(9,2)dW \\ &= 2 * 0.2334 \\ &= 0.4668\ ,\end{aligned}$$

similarly,

$$\begin{aligned}
\lambda_W(T=2) &= \int_W \lambda(W) \int_Y P(W|T=2,Y)\pi_W(Y)dYdW \\
&= \int_W 2\mathcal{N}(11,4)\mathcal{N}(11,2)dW \\
&= 2 * 0.3257 \\
&= 0.6514\ ,
\end{aligned}$$

and from (15), (16),

$$\begin{aligned}
\lambda_W(Y) &= \sum_T \left[\pi_W(T) \int_W \lambda(W)P(W|T,Y)dX\right] \\
&= 0.7273 \int_W 2\mathcal{N}(11,4)\mathcal{N}(-1+Y,1)dW + \\
&\quad 0.2727 \int_W 2\mathcal{N}(11,4)\mathcal{N}(1+Y,1)dW \\
&= 1.4546\mathcal{N}(12,5) + 0.5454\mathcal{N}(10,5)\ .
\end{aligned}$$

Therefore,

$$\begin{aligned}
\lambda(T) &= \lambda_C(T)\lambda_W(T) \\
&= [0.8\ 0.3].[0.4668\ 0.6514] \\
&= [0.37344\ 0.19542]\ , \\
\lambda(Y) &= \lambda_W(Y) \\
&= 1.4546\mathcal{N}(12,5) + 0.5454\mathcal{N}(10,5)\ .
\end{aligned}$$

Finally the beliefs of nodes T, Y can be computed as,

$$\begin{aligned}
BEL(T) &= \alpha\ \lambda(T)\pi(T) \\
&= \alpha\ [0.37344\ 0.19542].[0.5\ 0.5] \\
&= [0.65647\ 0.34353]\ , \\
BEL(Y) &= \alpha\ \lambda(Y)\pi(Y) \\
&= \alpha\ [1.4546\mathcal{N}(12,5) + 0.5454\mathcal{N}(10,5)] \\
&\quad \mathcal{N}(10,1) \\
&= \alpha\ [1.4546 * 0.1167\mathcal{N}(10.3333, 0.8333)] + \\
&\quad \alpha\ [0.5454 * 0.1629\mathcal{N}(10, 0.8333)] \\
&= 0.6564\mathcal{N}(10.3333, 0.8333) + \\
&\quad 0.3436\mathcal{N}(10, 0.8333)\ .
\end{aligned}$$

Now, the message passing algorithm provides the posterior distributions for all of hidden nodes T, Y, W. And since this is a poly tree model, the solution is exact.

Note that the presence of discrete parents for continuous variable makes the corresponding continuous messages necessarily a mixture distribution. Unfortunately, the number of mixture components in the message increases exponentially with the size of joint state space of the discrete parents. In order to scale the algorithm, one alternative is to combine or reduce the mixture components into smaller ones to trade off complexity against accuracy.

2.3 Complexity and scalability

The complexity of exact inference for a hybrid model is essentially determined by the size of the joint state space of all discrete parent nodes (i.e., interface nodes). It is easy to prove that, in a connected CLG, all discrete parents will end up in one clique with at least one continuous node Lerner (2002). Sometimes, even a CLG with very simple structure can give rise to an intractable clique tree. For example, the network shown in Figure 4 will have all of its discrete nodes in one clique, hence making the computations exponential in the size of the joint state space of all discrete nodes. If each discrete node has 10 states, then the resulting clique will have size 10^n, where n is the number of discrete nodes.

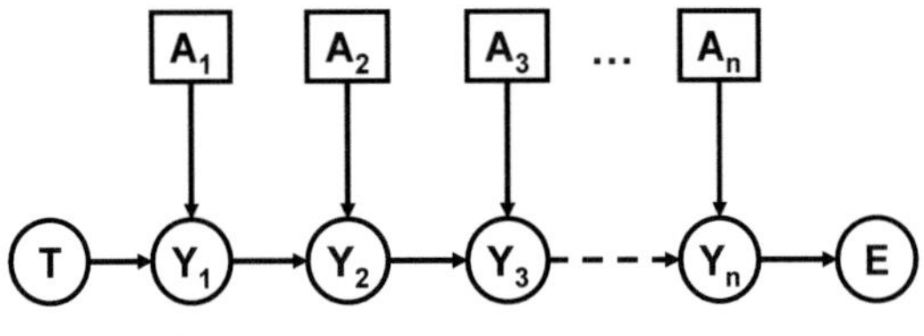

Fig. 4. A simple CLG that has an exponential clique tree: $A_1, A_2, ...A_n$ are discrete nodes, and $T, Y_1, Y_2, ..., Y_n, E$ are continuous nodes.

DMP-HBN has the same problem when exact inference is required. This is because for each state of a discrete parent node, its continuous child has to compute messages according to the function defined in the CPD. Therefore, messages sent by a continuous node with a hybrid CPD will be in the form of a Gaussian mixture in which the components are weighted by probabilities passed from its discrete parents. In particular, as shown in Equaion (10) and (12), $\pi(X)$ and $\lambda_X(U)$ are mixtures of Gaussians with the number of Gaussian components equal to the size of the state space of its discrete parent D. When a mixture message propagates to another continuous node with discrete parents, the message size will increase again exponentially. However, while JT has to deal with this intractability, DMP-HBN has the choice to approximate the original Gaussian mixture with a smaller number of components. In many cases, a Gaussian mixture with significantly fewer components can approximate the original density very well. Let us assume that $f(x)$ is the true density, and $\hat{f}(x)$ is the approximate Gaussian mixture. We use the following distance measure as the metric, called Normalized Integrated Square Error (NISE):

$$d = \frac{\int (f(x) - \hat{f}(x))^2 dx}{\int (f(x))^2 dx + \int (\hat{f}(x))^2 dx} .$$

An example shown in Figure 5 demonstrates a reasonable estimate using only 4 components to approximate a Gaussian mixture with 20 components ($\sqrt{d} < 3\%$). With a pre-defined error bound, Gaussian mixture reduction methods such as the ones proposed in Kuo-Chu Chang & Smith (2010) Schrempf et al. (2005) can be applied to find a good approximate mixture with a smaller number of components. It is straightforward to incorporate these methods into

DMP-HBN to make the algorithm scalable with an acceptable accuracy trade-off. However, it is non-trivial to estimate the overall inference error after the messages are compressed and propagated. In the next section, we will provide some performance results with numerical experiments to evaluate the algorithm under various situations.

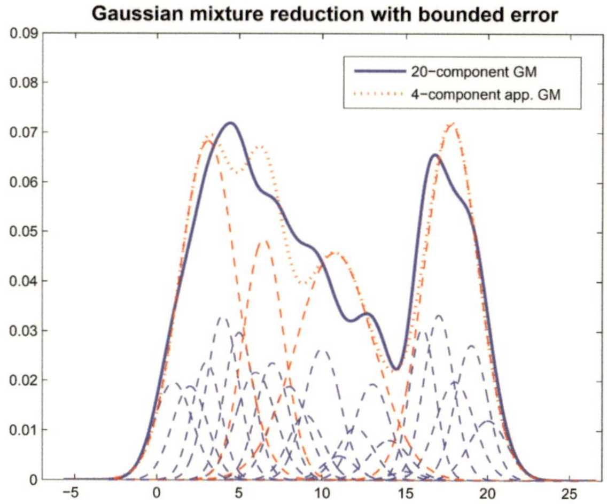

Fig. 5. Using a 4-component GM to approximate a 20-component GM with $\sqrt{d} < 3\%$.

3. Numerical experiments

Theoretically, DMP-HBN can provide exact results for a polytree CLG. For verification purpose, an example model called *Poly12CLG* as shown in Figure 6, was used for the experiment. Assume evidence is observed on leaf nodes E, and Z. With random observations, we conducted more than 30 independent experiments and compared DMP results with the ones obtained by the Junction Tree algorithm. The latter algorithm is considered to be the gold standard and the resulting solutions serve as the ground truth. All experiments show that DMP-HBN provides results identical to the ground truth.

We also conducted scalability tests of DMP algorithm using the same example model *Poly12CLG*. For many decision support applications, the variables of interest tend to be discrete, such as feature identification, entity classifications, or situation hypotheses. In our experiments, we first show how the assessments of hidden discrete nodes in a CLG are affected after collapsing the Gaussian mixture into a single Gaussian when passing messages. We use average absolute probability errors between two discrete distributions as the metric to evaluate the performance. In general, when a node of interest is relatively far away from the evidence, its posterior distribution would not deviate much from its prior. In that case, it is difficult to show the impact of the approximation on the inference error. So we purposely designed CPDs in *Poly12CLG* to move the true posterior probabilites away from its prior. Figure 7 shows the average and maximum errors of the approximate posterior probabilities for hidden discrete nodes V, A, L, B, H, and C, obtained after collapsing Gaussian mixtures into a single term over 100 Monte Carlo simulations. Average and maximum difference between the true posteriors and the priors over these 100 simulations are also shown in the figure for

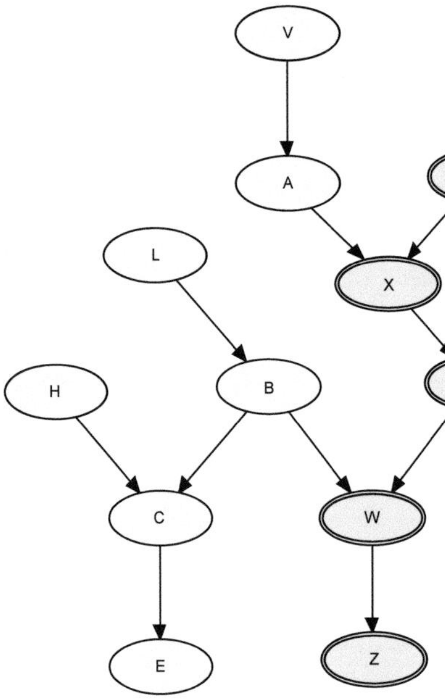

Fig. 6. An example polytree CLG model called Poly12CLG, consisting of 7 discrete nodes V, A, L, B, H, C, E and 5 continuous nodes U, X, Y, W, Z.

comparison. Figure 7(a) presents the estimate errors when collapsing π values only; Figure 7(b) shows the performance when collapsing λ messages only; and Figure 7 (c) displays the inference errors when collapsing both π values and λ messages whenever a mixture of Gaussians is present.

Notice that reducing the π value of a node does not affect the network "above" it because the π message is being sent downward in the network. Similarly, since a λ message is being sent upward, reducing a λ message will not affect the network "below" the node. For example in Figure 7(a), the posterior probabilities of V and A are exact, and in Figure 7(b), the estimates of $L, B, H,$ and C are also exact without inference error. When reducing both π values and λ messages, all posterior distributions are not exact any more. Results shown in Figure 7(c) suggest that the approximation errors diminish when the nodes are farther away from discrete parents. For example, the approximate errors for nodes $L, H,$ and C are very small. However, discrete parent nodes such as A, and B, are affected significantly. This is not surprising due to the relatively large approximation errors when collapsing a multi-modal Gaussian mixture into a single term. One way to achieve a desired accuracy is to specify a pre-defined error bound whenever we try to reduce a Gaussian mixture into one with fewer components. Although it is difficult to perform theoretical analysis of the total inference error after propagation, it is possible to obtain bounded error if the threshold used is small enough. Figure 8 demonstrates significantly better performance for the same model but with the normalized ISE of the reduced Gaussian mixture limited to less than 5% each time. As can be seen from the figure, the average and maximum errors for all nodes are well less than 1%.

Another example model called *Loop13CLG* (extended from *Poly12CLG*), shown in Figure 9, was used for numerical experimentation on a network with loops. Again, we assume that leaf

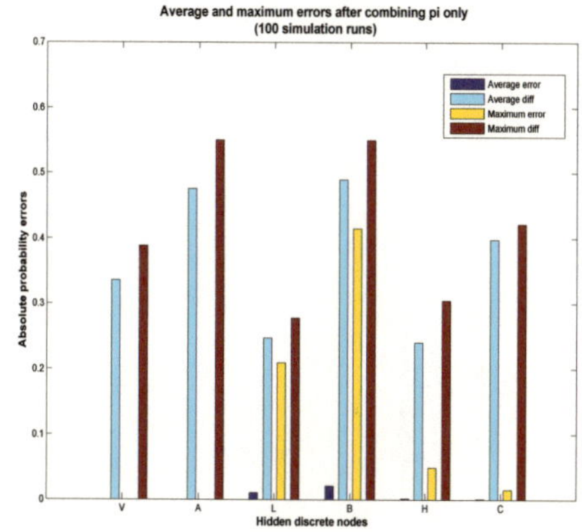

(a) combining π values only

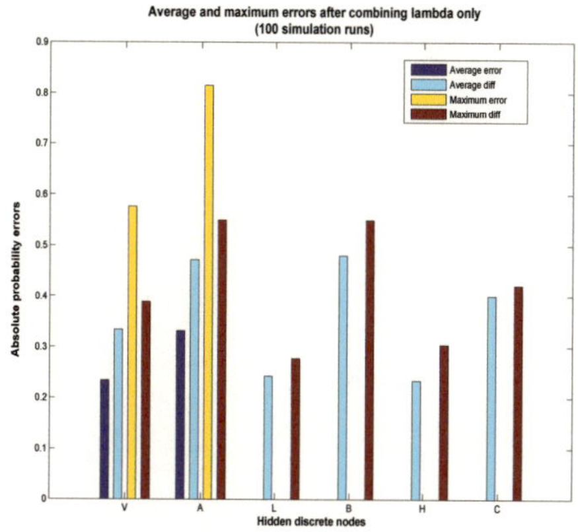

(b) combining λ messages only

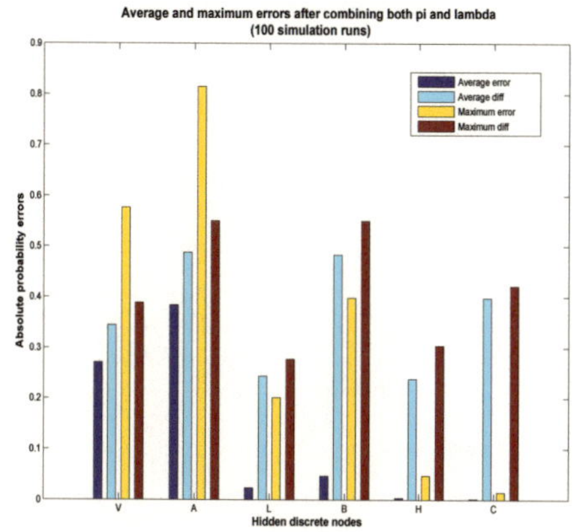

(c) combining both π values and λ messages

Fig. 7. Scalability test – performance loss after combining Gaussian mixture into one single Gaussian.

nodes E and Z are observable evidence nodes. With random observations, Figure 10 shows the average and maximum absolute errors of posterior probabilities for hidden discrete nodes over 100 Monte Carlo simulations. All simulation runs converge in about 11 iterations. As can be seen from the figure, average approximation errors caused by loopy propagation range from less than 1% to about 5% for hidden discrete nodes.

We also tested DMP with some other networks with randomly pre-defined CPDs. All simulation results suggest that the estimation errors reduce significantly as the node is farther away from the discrete parent nodes.

4. Most probable explanation for hybrid Bayesian networks

In addition to computing the posterior distributions for hidden variables in Bayesian networks, one other important inference task is to find the most probable explanation (MPE).

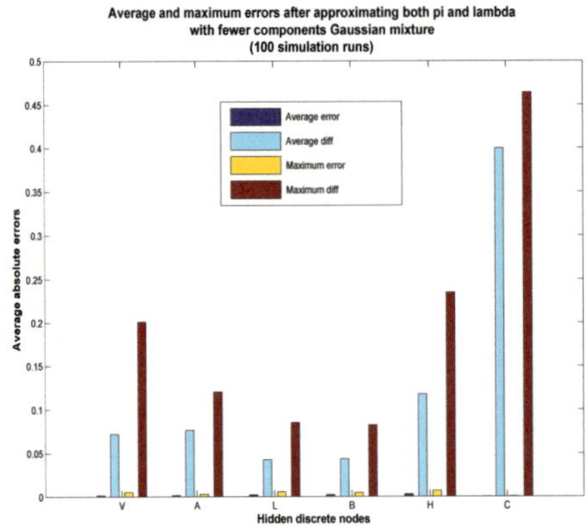

Fig. 8. Accurate estimates of the posterior probabilities resulted by limiting approximation error (< 5%) each time when reducing message with fewer components Gaussian mixture.

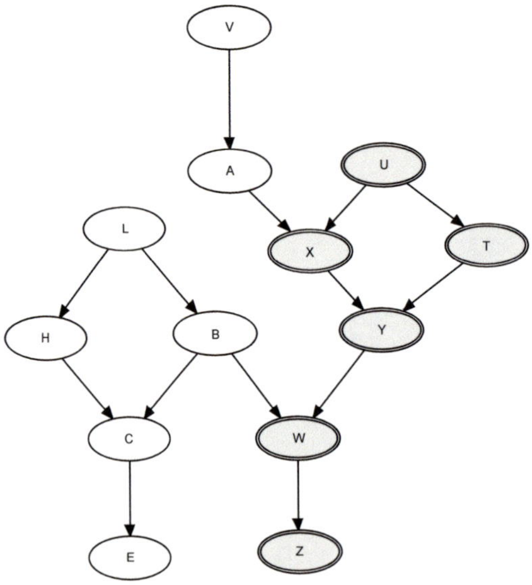

Fig. 9. Loop13CLG – an example CLG model with multiple loops, consisting of 7 discrete nodes V, A, L, B, H, C, E and 6 continuous nodes U, X, TY, W, Z.

MPE provides the most likely configurations to explain away the evidence and helps to manage hypotheses for decision making. In recent years, researchers have proposed a few methods to find the MPE for discrete Bayesian networks. However, finding the MPE for hybrid networks remains challenging. In the following sections, we will briefly describe an up-to-date method to find the MPE in hybrid BNs based on max-product clique tree algorithm.

Let $\mathcal{X}$ represents the full set of variables in a Bayesian network, and $\mathbf{E}$ as a subset of $\mathcal{X}$ containing variables observed, known as evidence. The MPE is the joint assignment of

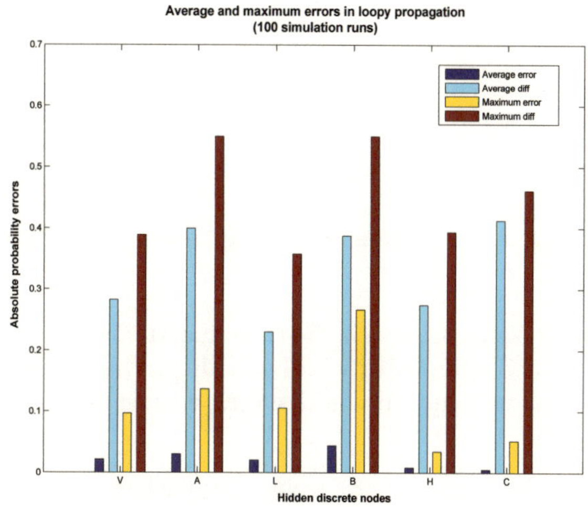

Fig. 10. Performance test with loopy CLG model.

$\mathbf{W} = \mathcal{X} \setminus \mathbf{E}$(subset of all hidden variables) such that:

$$\text{MPE } P(\mathbf{W}|\mathbf{E} = \mathbf{e}) = \arg\max_{\mathbf{w}} \; P(\mathbf{W} = \mathbf{w}|\mathbf{E} = \mathbf{e}) \tag{19}$$

where $\arg\max_x \; f(x)$ represents the value of x for which $f(x)$ is maximal.

Note that we have to look at the joint assignment to maximize the joint probability. Individually most likely values of variables that maximize their marginal probabilities are not necessarily part of the MPE. A very simple example is given below to demonstrate this point. Let us look at the BN model consisting of only 3 nodes ($D, E,$ and, F), shown in Figure 11, where $D, E,$ and F are binary discrete random variables with the CPDs listed in the figure.

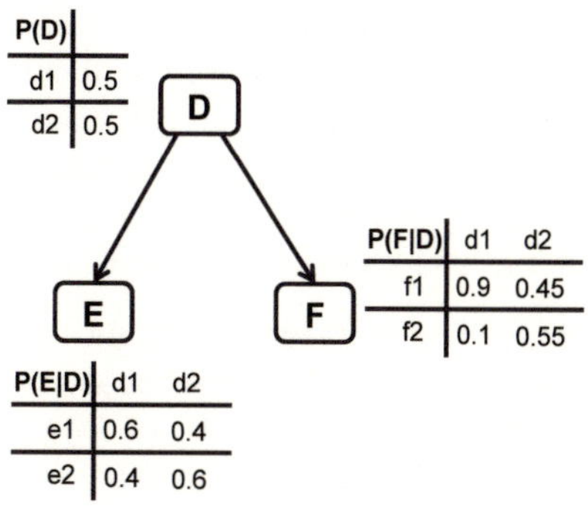

Fig. 11. A simple Bayesian network model consisting of 3 binary discrete nodes ($D, E,$ and, F).

Now let us assume that E is observed as $e2$. It is easy to show that

$$P(D|E = e2) = \begin{bmatrix} d1 : 0.4 \\ d2 : 0.6 \end{bmatrix}, \; P(F|E = e2) = \begin{bmatrix} f1 : 0.63 \\ f2 : 0.37 \end{bmatrix},$$

and

$$P(F|D, E = e2) = P(F|D) = \begin{bmatrix} & \text{d1} & \text{d2} \\ \hline \text{f1} & 0.9 & 0.45 \\ \text{f2} & 0.1 & 0.55 \end{bmatrix}.$$

Therefore

$$P(D, F|E = e2) = P(F|D, E = e2) \times P(D|E = e2) = \begin{bmatrix} & \text{d1} & \text{d2} \\ \hline \text{f1} & 0.36 & 0.27 \\ \text{f2} & 0.04 & 0.33 \end{bmatrix}.$$

From the joint probability distribution, it is clear that the MPE of $E = e2$ is the configuration of $D = d1, F = f1$. If we choose the MPE by individually picking up the values with maximal marginal probabilities, one will end up with a wrong answer as $D = d2, F = f1$, which is obviously not the true MPE.

Theoretically, to compute the maximal joint probability, we have

$$\max_{\mathbf{W}} P(\mathbf{W}|\mathbf{E} = \mathbf{e}) = \max_{W_i} \prod_{i=1}^{n} P(W_i|\text{Pa}(W_i), \mathbf{E} = \mathbf{e}) \tag{20}$$

where $W_i(i = 1, 2, \ldots, n)$ are all of the hidden variables in $\mathbf{W}$ (with the total number of variables in $\mathbf{W}$ being n), and $\text{Pa}(W_i)$ are the parents of node W_i. Clique Tree algorithm has been used in computing the MPE Koller & Friedman (2009), where one needs to replace the marginalization operation with maximization operation for each potential. In this paper, we call this method the *max-product clique tree algorithm*. And accordingly, the potentials in the clique tree are called *max-potentials*.

4.1 Max-calibration of the clique tree for discrete Bayesian networks

The standard clique tree algorithm is a generalization of the variable elimination method for Bayesian network inference. It first transforms the original Bayesian network into a clique tree, which is a undirected poly-tree with cliques serving as nodes in the tree. Each clique is a joint state space of more than one variables, associating with a function called potential. Once a root clique is chosen, one needs to conduct a round trip message propagations in order to have each clique updated by the given evidence. The message propagation from leaf nodes to the root along the path is called upstreaming, also known as collecting evidence; while the opposite is called downstreaming, also known as distributing evidence.

In the process of message propagation between two cliques, a standard protocol is applied. Let us assume that two cliques $\mathbf{C}_i, \mathbf{C}_j$ are neighbors in a clique tree, and separator $\mathbf{S}_{ij}$ is associated with the edge between $\mathbf{C}_i, \mathbf{C}_j$. Potentials for $\mathbf{C}_i, \mathbf{C}_j$ and $\mathbf{S}_{ij}$ are $\phi(\mathbf{C}_i), \phi(\mathbf{C}_j)$ and $\phi(\mathbf{S}_{ij})$ respectively. Sending message from $\mathbf{C}_i$ to $\mathbf{C}_j$ along the separator $\mathbf{S}_{ij}$ follows the message passing protocol presented in Table 1. The sending process is also known as absorption, namely, clique $\mathbf{C}_j$ absorbs information from clique $\mathbf{C}_i$ via their separator $\mathbf{S}_{ij}$.

Note that in Table 1, the first step of message propagation is to marginalizing out the variables in $\mathbf{C}_i$ but not in $\mathbf{C}_j$, so only variables in the separator are left. This is why traditional clique tree algorithm is sometimes called sum-product clique tree method due to this summing out

1. Let $\phi(\mathbf{S}_{ij})' = \sum_{\mathbf{C}_i \backslash \mathbf{S}_{ij}} \phi(\mathbf{C}_i)$, — marginalizing the potential $\phi(\mathbf{C}_i)$ onto the domain of separator $\phi(\mathbf{S}_{ij})$, i.e., projecting it to the domain of separator.
2. Let $\mathcal{L}(\mathbf{S}_{ij}) = \frac{\phi(\mathbf{S}_{ij})'}{\phi(\mathbf{S}_{ij})}$, — dividing the new potential of separator $\phi(\mathbf{S}_{ij})$ by its old one. The ratio $\mathcal{L}(\mathbf{S}_{ij})$ is served as information ratio, also called "likelihood ratio", to update information by filtering out the redundant part.
3. Let $\phi(\mathbf{S}_{ij}) = \phi(\mathbf{S}_{ij})'$, — storing the new potential of the separator for next round message passing.
4. Let $\phi(\mathbf{C}_j) = \phi(\mathbf{C}_j) * \mathcal{L}(\mathbf{S}_{ij})$, — multiplying information ratio from the separator to update potential of $\phi(\mathbf{C}_j)$.

Table 1. Message passing protocol in standard clique tree algorithm

operation. In max-calibration of a clique tree, maximizing replaces marginalizing, while all other operations remain the same in the protocol.

In discrete case, for MPE, it is straightforward to maximize out variables from the joint state distribution. Suppose that we have a joint probability distribution of two binary discrete random variables D, T(states of D, T are $d1, d2, t1$ and $t2$ respectively), shown as below:

$$P(D,T) = \left[\begin{array}{c|cc} & \text{t1} & \text{t2} \\ \hline \text{d1} & 0.32 & 0.16 \\ \text{d2} & 0.39 & 0.13 \end{array}\right].$$

To maximize out T, we have

$$\max_T P(D,T) = \max_T \left[\begin{array}{c|cc} & \text{t1} & \text{t2} \\ \hline \text{d1} & 0.32 & 0.16 \\ \text{d2} & 0.39 & 0.13 \end{array}\right] = \left[\begin{array}{c|c} \text{d1} & 0.32 \\ \text{d2} & 0.39 \end{array}\right]$$

Similarly, if we want to maximize out D from the joint distribution of D, T, it will be:

$$\max_D P(D,T) = \max_D \left[\begin{array}{c|cc} & \text{t1} & \text{t2} \\ \hline \text{d1} & 0.32 & 0.16 \\ \text{d2} & 0.39 & 0.13 \end{array}\right] = \left[\begin{array}{c|c} \text{t1} & 0.39 \\ \text{t2} & 0.16 \end{array}\right]$$

In principle, maximizing out one variable from a joint discrete space returns the marginal maximums in the original joint probabilities along the dimension of this particular variable being maximized over, for all of the configurations of the remaining variables.

With the maximizing substituted in the message propagation protocol, the clique tree will be max-calibrated after conducting the same upstream and downstream message propagations. Then each clique will be updated with the max-potential. After the max-calibration, further maximizations on individual cliques can provide the MPE of all hidden variables. The proofs are very similar to the proofs of the standard clique tree algorithm Jensen (1996) Dawid (1992).

4.2 Finding the MPE for a hybrid Bayesian network using max-product clique tree algorithm

For a hybrid Bayesian network, its clique tree contains at least one hybrid clique, in which both discrete and continuous variables are involved. If we can find a way to conduct maximizing operations for the hybrid clique, we can then apply the max-calibration process similarly to find the MPE for hybrid BNs.

Let us first take a close look at the hybrid joint space. Without loss of generality, we assume that the continuous variables in the hybrid space are Gaussians. For arbitrary density, theoretically, it is well known that a Gaussian mixture can be used to approximate the original density in any desirable accuracy with sufficient number of components. A simple example of the hybrid space, consisting of one binary discrete variable D with states $d1, d2$, and one scalar Gaussian variable X, is used for demonstration. Assuming that the hybrid joint density is

$$P(D,X) = \left[\begin{array}{c|l} & x \\ \hline d1 & 0.2\mathcal{N}(x;1,0.1) \\ d2 & 0.8\mathcal{N}(x;3,3) \end{array}\right],$$

where $\mathcal{N}(x;u,\sigma^2)$ represents a scalar Gaussian density with mean u, and variance σ^2, and x is a real number. Note that $P(D,X)$ is not a conditional density, nor is a Gaussian mixture, but a hybrid joint density. For example, the joint density for $D = d1, X = 0.5$ is $0.2 * \mathcal{N}(0.5;1,0.1) = 0.325$. If we sum out D, we then can obtain the marginal distribution of X as the linear combination of two Gaussians with weights as $0.2, 0.8$ respectively, which is indeed a Gaussian mixture:

$$P(X) = 0.2\mathcal{N}(x;1,0.1) + 0.8\mathcal{N}(x;3,3).$$

Next, let us see how to maximize one variable from the hybrid joint space using this example. The resulting function after maximizing over some variables is mapped onto the space of the remaining variables. If the variable being maximized out is D from $P(D,X)$, by applying the maximizing rule, we have

$$P(X)^{max} = \max_D P(D,X) = max[0.2\mathcal{N}(x;1,0.1), 0.8\mathcal{N}(x;3,3)],$$

where $P(X)^{max}$ is called the marginal maximum function of X. This is basically a function of x with the values as either $0.2\mathcal{N}(x;1,0.1)$, or $0.8\mathcal{N}(x;3,3)$, whichever is bigger for a given x. As shown in Figure 12, the max of two Gaussians is not a Gaussian mixture. However, because of the closed form of Gaussian density, we can deterministically conclude that the peak of this function is certainly located at one of the mean values among all Gaussian components. Proof is omitted due to its obviousness.

Now let us turn to maximizing out the continuous variable X from the hybrid density $P(D,X)$. Again, by applying the maximizing rule, it is easy to obtain,

$$P(D)^{max} = \max_X P(D,X) = \left[\begin{array}{c|l} d1 & 0.2\mathcal{N}(x=1;1,0.1) \\ d2 & 0.8\mathcal{N}(x=3;3,3) \end{array}\right] = \left[\begin{array}{c|l} d1 & 0.2523 \\ d2 & 0.1843 \end{array}\right],$$

which are the peak points of densities for each weighted Gaussian component given each state of the discrete variable, respectively. In the case of Gaussian density, the peak point is obviously located at the mean, namely, $max[\mathcal{N}(x;u,\sigma^2)] = \mathcal{N}(x=u;u,\sigma^2)$.

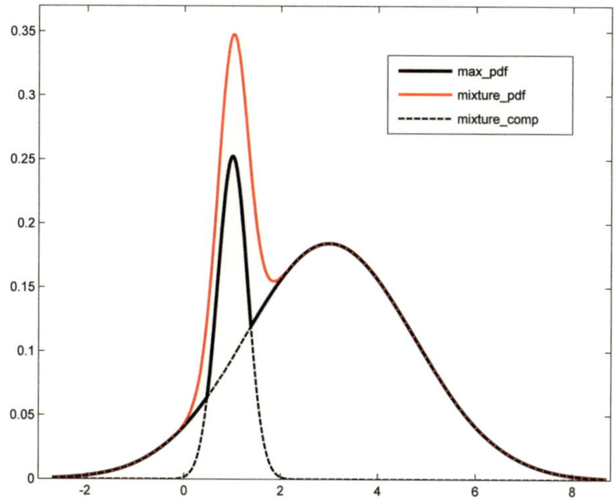

Fig. 12. Maximizing out D from $P(D,X)$ — the resulting function is the max of weighted Gaussian components. In the figure, the red line represents the density of the Gaussian mixture, the bolded black line represents the resulting max function, and the dashed black line shows the original Gaussian components.

Accordingly, the MPE of $P(D,X)$ can be obtained by further examining the value that maximizes the marginal maximum function for each variable. Then,

$$\text{MPE } P(D,X) = \{\arg\max_D P(D)^{max}, \arg\max_X P(X)^{max}\} = \{D = d1, x = 1\}.$$

At this point, we know how to maximize out variables from both discrete joint space and hybrid space. We still need to know how to maximize out variable from continuous joint space in order to conduct max-calibration for hybrid model. Again, we assume that the continuous variables are Gaussian. Maximizing out a continuous variable from continuous joint space is equivalent to having the value of this variable being its marginal mean and then substituting it into the original joint density function. Let us use a two-dimension Gaussian density to explain the operation. For the sake of simplicity, we assume the two Gaussians X, Y in different dimensions are independent of each other. Namely,

$$P(X,Y) = \mathcal{N}\left(\begin{bmatrix} x \\ y \end{bmatrix}; \begin{bmatrix} u_x \\ u_y \end{bmatrix}, \begin{bmatrix} \sigma_x^2 & 0 \\ 0 & \sigma_y^2 \end{bmatrix}\right),$$

where $\begin{bmatrix} u_x \\ u_y \end{bmatrix}$ is the mean vector, and $\begin{bmatrix} \sigma_x^2 & 0 \\ 0 & \sigma_y^2 \end{bmatrix}$ is the covariance matrix. Maximizing out Y from $P(X,Y)$ to obtain the marginal maximum function of X is carried out as below,

$$\begin{aligned} P(X)^{max} &= \max_Y P(X,Y) = \arg\max_Y \frac{1}{2\pi\sigma_x\sigma_y} \exp\left(-\frac{1}{2}\left[\frac{(x-u_x)^2}{\sigma_x^2} + \frac{(y-u_y)^2}{\sigma_y^2}\right]\right) \\ &= \frac{1}{\sqrt{2\pi}\sigma_y} \mathcal{N}(x; u_x, \sigma_x^2). \end{aligned}$$

Given a clique tree of the hybrid model, a strong root of the clique tree, and evidence, this algorithm returns the MPE of the evidence for the original hybrid Bayesian network.

1. Sending messages from leaf cliques to the strong root clique: message passing between cliques follows the protocol shown in Table 1 except using maximizing to replace marginalizing.
2. After the strong root clique receives all messages, sending back messages to all leaf cliques: message passing between cliques follows the protocol shown in Table 1 except using maximizing to replace marginalizing.
3. Conducting further maximizing operation on each clique to obtain the marginal maximum function for each hidden variable, then choosing the value of variable that maximizes its marginal maximum function. Those values together compose the MPE.

Table 2. Hybrid max-product clique tree algorithm to find the MPE for hybrid Bayesian networks

Similar derivation can be done for higher-dimension cases, and/or with dependent variables.

4.3 Division and multiplication between functions

In message passing protocol, shown in Table 1, we also note that division and multiplication operations need to be defined for hybrid models. In this case, the only difference from the discrete case is that how to apply continuous functions in these operations. The result of functional division or multiplication may not have a closed form, but could be computed for any given value of argument variable numerically. And for Gaussian densities, the peak value of the resulting function can be obtained deterministically.

4.4 Hybrid max-product clique tree algorithm (HMP-CT)

Now we are ready to present the hybrid max-product cliquet tree algorithm (HMP-CT) for finding the MPE in hybrid Bayesian networks in Table 2.

4.5 Numerical example - finding Hybird MPE

In this section, we use a simple hybrid model to demonstrate HMP-CT. With the model shown in Figure 11, we change the node F to be a Gaussian variable and all other parameters and network structure remain the same. The hybrid model with its new CPD is shown in Figure 13(a), where the ellipse is used to represent continuous variable.

There are only two cliques in the corresponding clique tree of the model, shown in Figure 13(b), in which the clique $\{D, F\}$ is the strong root. Assuming that the observed evidence $E = e2$, let us follow the algorithm described in Table 2 to find the MPE configuration of the hidden nodes D, F.

First, the initial potentials of these two cliques are

$$\phi(D,E) = \left[\begin{array}{c|c} & \text{E=e2} \\ \hline d1 & 0.4 \\ d2 & 0.6 \end{array}\right], \quad \phi(D,F) = \left[\begin{array}{c|c} & f \\ \hline d1 & 0.5\mathcal{N}(f;1,0.5) \\ d2 & 0.5\mathcal{N}(f;3,2) \end{array}\right].$$

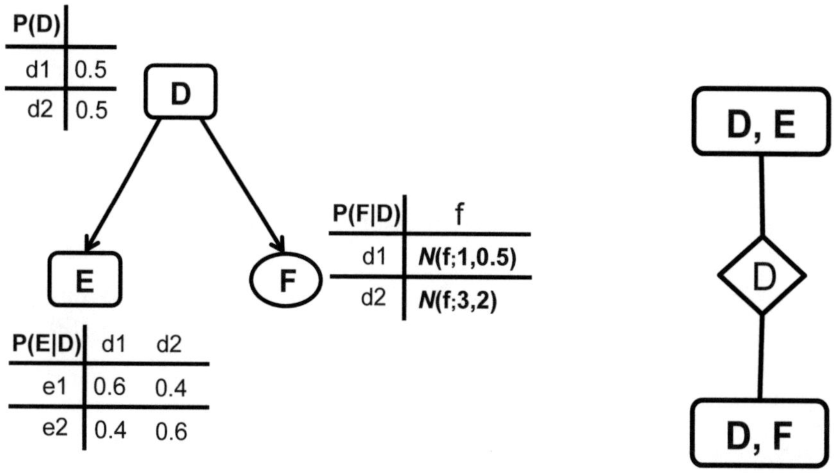

(a) network structure and its CPDs (b) The corresponding clique tree

Fig. 13. A simple hybird Bayesian network model consisting of 2 binary discrete nodes (D, E) and one Gaussian variable (F).

And the potential of the only separator D is uniformly initialized as $\phi(D) = \begin{bmatrix} 1 \\ 1 \end{bmatrix}$. Since the strong root is $\{D, F\}$, the upstreaming message passing is then from the clique $\{D, E\}$ to $\{D, F\}$. We have

$$\phi(D)' = P(D)^{max} = \max_E \phi(D, E) = \left[\begin{array}{c|c} d1 & 0.4 \\ d2 & 0.6 \end{array}\right].$$

The updated potential of D, F is,

$$\phi(D, F) \leftarrow \phi(D, F) \times \frac{\phi(D)'}{\phi(D)} = \left[\begin{array}{c|c} & f \\ \hline d1 & 0.2\mathcal{N}(f; 1, 0.5) \\ d2 & 0.3\mathcal{N}(f; 3, 2) \end{array}\right].$$

Also we need to update the potential of the separator D to be $\phi(D) = \phi(D)' = \left[\begin{array}{c|c} d1 & 0.4 \\ d2 & 0.6 \end{array}\right]$. Now sending back the message from the root $\{D, F\}$ to the leaf $\{D, E\}$, we have

$$\phi(D)' = \max_F \phi(D, F) = \left[\begin{array}{c|c} d1 & 0.2\mathcal{N}(f = 1; 1, 0.5) \\ d2 & 0.3\mathcal{N}(f = 3; 3, 2) \end{array}\right] = \left[\begin{array}{c|c} d1 & 0.1128 \\ d2 & 0.0846 \end{array}\right].$$

Again, the potential of D, E is updated as,

$$\phi(D, E) \leftarrow \phi(D, E) \times \frac{\phi(D)'}{\phi(D)} = \left[\begin{array}{c|c} & \text{E=e2} \\ \hline d1 & 0.1128 \\ d2 & 0.0846 \end{array}\right].$$

Now the max-calibration of the clique tree is complete. By further maximizing the potentials onto each hidden variable, we have

$$\phi(D)^{max} = \max_E \phi(D,E) = \left[\begin{array}{c|c} d1 & 0.1128 \\ d2 & 0.0846 \end{array}\right],$$

and

$$\phi(F)^{max} = \max_D \phi(D,F) = max(0.2\mathcal{N}(f;1,0.5), 0.3\mathcal{N}(f;3,2)) = max(0.1596, 0.0598) = 0.1596,$$

located at $f = 1$. Therefore, the MPE of $E = e2$ is $\{D = d1, F = 1\}$. From the joint posterior distribution $P(D,F|E = e2)$, the peak value of joint density associated with the MPE is 0.2257.

5. Summary

In this chapter, we presented a new inference algorithm called DMP-HBN to represent probabilistic messages in the form of Gaussian mixture when continuous variables are involved and allow exchanging messages between discrete and continuous variables directly. This new algorithm provides an alternative for probabilistic inference in hybrid Bayesian networks. It provides full density estimates for continuous variables and can be extended with unscented transformation Sun & Chang (2007a) for the general hybrid models with nonlinear and/or non-Gaussian distributions. Since DMP-HBN is a distributed algorithm utilizing only local information, there is no need to transform the network structure as required by the Junction Tree algorithm. Compared to our previous works in Sun & Chang (2007b), Sun & Chang (2009), that need to partition the hybrid model into different network segments, and then conduct message passing separately, DMP-BN can exchange messages directly between discrete and continuous variables within an unified framework. In addition, the algorithm does not require prior knowledge of the global network topology which could be changing dynamically. This is a major advantage of the algorithm and is particularly important to ensure scalable and reliable message exchanges in a large information network where computations are done locally.

As shown in the empirical simulation results, DMP-HBN is scalable with a performance tradeoff of losing some accuracy. For many decision support applications, we are mainly interested in hidden discrete variables such as entity classifications or high level situation hypotheses. The experimental results show that the estimation errors of the hidden discrete variables depend on the network topology and are relatively modest, especially when the variables of interest are far away from the discrete parent nodes. Theoretically, it is non-trivial to estimate the overall performance bounds quantitatively due to message compressing and propagation. Even though we can have the error bounded each time when we approximate the original Gaussian mixture with less number of components, it is theoretically difficult to estimate the total error after we propagate the approximate messages multiple times. Similar problem exists in filtering for stochastic dynamic systems. This points to an important and very interesting topic for future research.

In addition to the inference task of calculating posterior distributions, finding the MPE is another important type of inference and it has a number of real-life applications in decision support. In the chapter, we introduced and descibed in detail a hybrid max-calibration clique tree algorithm, called HMP-CT, to find the MPE for hybrid Bayesian networks. We derived all of required operations in the calibration process. Different from the standard sum-product clique tree algorithm, HMP-CT maximizes out variables from the clique potentials instead of marginalizing.

As mentioned in Section 4.3, division and multiplication in message propagation process for hybrid model require functional operations. Further investigations are needed in order to find the better representations of the resulting functions to save computations. In the process, what we need is to obtain the locations (values of variables), where maximize the resulting functions.

To our best knowledge, little research has been done for finding the MPE in hybrid BN models. On the other hand, it is almost inevitable to have continuous variables involved when modeling a real-life problem. It is especially useful to have the MPE for managing multiple most likely hypotheses in many decision support systems. Also, finding the MPE is essentially a global searching problem as to find the maximum. For a genearal optimization problem, if we can decompose the joint state space and model the cost function by a BN-like structure, we can then apply max-calibration algorithm to solve it.

Similarly and with the obvious proof, min-calibration clique tree algorithm can provide the least probable explanation (LPE). In some interesting domains, the LPE is very useful. For example, in prediction market, we always need to know the minimum possible asset of a trader in case that some random events happen to occur that are against this trader's bet.

6. References

Chang, K. C. & Sun, W. (2010). Scalable fusion with mixture distributions in sensor networks, *2010 11th International Conference on Control Automation Robotics & Vision (ICARCV)*, IEEE, pp. 1251–1256.

Charniak, E. (1991). Bayesian networks without tears: making bayesian networks more accessible to the probabilistically unsophisticated, *AI Magazine* 12(4): 50–63.

Cheng, J. & Druzdzel, M. J. (2000). AIS-BN: an adaptive importance sampling algorithm for evidential reasoning in large bayesian networks, *Journal of Artificial Intelligence Research* 13: 155–188.

Cobb, B. R. & Shenoy, P. P. (2006). Inference in hybrid bayesian networks with mixtures of truncated exponentials, *International Journal of Approximate Reasoning* 41: 257–286.

Dawid, A. (1992). Application of a general propagation algorithm for probabilistic expert systems, *Statistics and Computing* 2(1): 25–36.

Fung, R. & Chang, K. C. (1989). Weighting and integrating evidence for stochastic simulation in bayesian networks, *Uncertainty in Artificial Intelligence 5*, Elsevier Science Publishing Company, Inc., New York, pp. 209–219.

Gamerman, D. & Lopes, H. F. (2006). *Markov chain Monte Carlo: stochastic simulation for Bayesian inference*, CRC Press.

Gilks, W. R., Richardson, S. & Spiegelhalter, D. J. (1996). *Markov chain Monte Carlo in practice*, CRC Press.

Jensen, F. (1996). *An Introduction to Bayesian Networks*, Springer-Verlag, New York.

Julier, S. J. (2002). The scaled unscented transformation, *Proceedings of the American Control Conference*, Vol. 6, pp. 4555–4559.

Koller, D. & Friedman, N. (2009). *Probabilistic Graphical Models: Principles and Techniques*, MIT Press, Cambridge, Mass.

Koller, D., Lerner, U. & Angelov, D. (1999). A general algorithm for approximate inference and its application to hybrid bayes nets, *Proceedings of the 15th Annual Conference on Uncertainty in Artificial Intelligence (UAI)*, pp. 324–333.

Kuo-Chu Chang, H. C. & Smith, C. (2010). Constraint optimized weight adaptation for gaussian mixture reduction, *Proceeding of SPIE Conference on Defense, Security, and Sensing*, Orlando,FL.

Lauritzen, S. (1992). Propagation of probabilities, means amd variances in mixed graphical association models, *JASA* 87(420): 1098–1108.

Lerner, U. N. (2002). *Hybrid Bayesian Networks for Reasoning about Complex Systems*, Stanford University.

Murphy, K., Weiss, Y. & Jordan, M. (1999). Loopy belief propagation for approximate inference:an empirical study, *Proceedings of the Fifteenth Conference on Uncertainty in Artificial Intelligence.*

Neapolitan, R. (1990). *Probabilistic Reasoning in Expert Systems*, John Wiley & Sons, New York.

Pearl, J. (1988). *Probabilistic Reasoning in Intelligent Systems: Networks of Plausible Inference*, Morgan Kauffman, San Mateo.

Schrempf, O. C., Feiermann, O. & Hanebeck, U. D. (2005). Optimal mixture approximation of the product of mixtures, *Proceedings of the 8th International Conference on Information Fusion (Fusion 2005)*, Vol. 1, Philadelphia, Pennsylvania, pp. 85–92.

Shachter, R. D. & Peot, M. A. (1999). Simulation approaches to general probabilistic inference on belief networks, *Proceedings of the Conference on Uncertainty in Artificial Intelligence*, Vol. 5.

Shenoy, P. (2006). Inference in hybrid bayesian networks using mixtures of gaussians, *Proceedings of the Twenty-Second Conference Annual Conference on Uncertainty in Artificial Intelligence (UAI-06)*, AUAI Press, Arlington, Virginia, pp. 428–436.

Sudderth, E. B., Ihler, A. T., Freeman, W. T. & Willsky, A. S. (2003). Nonparametric belief propagation and facial appearance estimation, *Computer Vision and Pattern Recognition*, Vol. 11, p. 1.

Sun, W. & Chang, K. (2007a). Unscented message passing for arbitrary continuous bayesian networks, *Proceedings of the 22nd AAAI Conference on Artificial Intelligence*, Vancouver, Canada.

Sun, W. & Chang, K. (2009). Message passing for hybrid bayesian networks: Representation, propagation, and integration, *IEEE Transaction on Aerospace and Electronic Systems* 45: 1525–1537.

Sun, W. & Chang, K. C. (2007b). Hybrid message passing for general mixed bayesian networks, *Proceedings of the 10th International Conference on Information Fusion*, Quebec, Canada.

Wainwright, M. J. & Jordan, M. I. (2008). *Graphical Models, Exponential Families, and Variational Inference*, Now Publishers Inc.

Yuan, C. & Druzdzel, M. J. (2006). Hybrid loopy belief propagation, *Proceedings of the third European Workshop on Probabilistic Graphical Models*, pp. 317–324.

Yuan, C. & Druzdzel, M. J. (2007). Generalized evidence pre-propagated importance sampling for hybrid bayesian networks, *Proceedings of the 22nd national conference on Artificial intelligence - Volume 2*, AAAI Press, p. 1296–1302. ACM ID: 1619853.

6

BN Applications in Operational Risk Analysis: Scope, Limitations and Methodological Requirements

Paolo Trucco[1] and Maria Chiara Leva[2]
[1]*Politecnico di Milano*
[2]*Trinity College*
[1]*Italy*
[2]*Ireland*

1. Introduction

Modern societies, due to their intrinsic complexity, are strongly dependent on critical resources and even more vulnerable to uncertain conditions. Despite the ability of controlling technical processes has increased over the past century, several different external and internal factors continue to affect the overall performance and sustainability of modern socio-technical systems. Globalisation, technology innovation and the organisational complexity of several actors are some of the major sources of uncertainty alongside the political context.

Emerging risks, also sometimes called global risks, are large-scale events or circumstances that arise from global trends; are beyond any particular party's capacity to control; and may have impacts not only on the organisation but also on multiple parties across geographic borders, industries, and/or sectors, in ways difficult to imagine today.

Moreover, modern societies are sustained and shaped by large socio-technical systems, where technology is deeply integrated with the human element and the organisational dimension. The identification and management of the wide spectrum of risks affecting such system of systems require new approaches and methods able to properly model and account for the growing complexity and dynamic interconnectedness of the modern world.

In this perspective, many organisations have deployed risk management programmes to identify, assess, and manage risks, using techniques such as risk assessment, scenario analysis, and stress testing as a basis for determining response strategies aligned with the entity's objectives, risk appetite and tolerance.

The recent world economic crisis pointed out two important lessons in the risk management field. The first is related to the continuous attempt of academics and practitioners to research for new approaches to predicting emerging risks and possible disaster scenarios that can irremediably affect operations or business viability. In the recent years top management, especially in the financial sector, paid more attention into sophisticated techniques, able to assure a limited exposure to specific risks, but that, on the other hand, opened to a wider exposure to correlated or systemic risks. As evidence, this approach made companies and the entire global economy more vulnerable than ever (Taleb et al., 2009). The second lesson learnt is that industrial organizations are facing highly differentiated risks, by types and scale, than the ones faced by the financial sector. An example is given by some automotive companies pushed down in the market by the same risks they had assumed for twenty years by generating profits only from energy not efficient vehicles (Kaplan et al., 2010). Moreover, risks affecting customers, employees and long-term viability of the business model are claiming for a wider understanding of risk nature and related correlations. Instead of designing more sophisticated tools to anticipate such catastrophic events, should be urgent a deeper understanding of the nature of operational risks and the development of a more integrated way to address these risks among the entire enterprise levels and entities, in order to foster its resilience and sustainability (Silvestri, 2010).

2. Evolutions in operational risk analysis and management

The category of "operational risk" was conceived as a composite term for a wide variety of organizational and behavioural risk factors which were traditionally excluded from formal definitions of market and credit risk (Power, 1993). Operational risk is much more than risks related to operations; in fact operations risk is a subset of the operational risks, only including risks related to the production process and planning (Samad-Khan A., 2008).

However the growing attention to operational risks is putting into light that new effort is needed not to merely re-label or codify a well established set of risk factors, but to develop a coherent new body of knowledge for the effective management of a complex phenomenon. The challenge calls for a real integration between professional and scientific contributions and perspectives (Power, 1993; Abbott, 1988).

A still widely used definition of operational risk was firstly proposed in the financial sector: "the risk of direct or indirect loss resulting from inadequate or failed internal processes, people and systems or from external events" (BCBS, 2001). The apparent aim of this definition is to give operational risk a clear and actionable focus on losses, although this definition still leaves open a range of operational risk attributes. For example, in the transportation industry operational risk management was defined by Beroggi and Wallace (1994) as "a decision logic to support individual or group-level reasoning processes in risky, time constrained situations when the need for plan revision arises". Here, the authors focused on the relevance of operational risk management for decision making, but at the same time reduced its scope to real-time or tactical decisions.

The potential targets exposed to operational risks can be identified by considering which company's entities are affected by uncertain events; indeed, operational risk results from the potential disruptions in the core operating, manufacturing or processing capabilities of a generic organisation.

In conclusion, operational risks can be defined as those interactions between an uncertain event and internal organisation's processes and/or resources, with the potential of influencing the core capabilities and resulting in a value variation over a time horizon (Silvestri et al., 2009; Trucco et al., 2010).

2.1 Classification of Operational Risks

The evolution towards an integrated approach to Operational Risk Management (ORM) raised the need of a comprehensive risk classification. To this end a basic classification of enterprise risks can be firstly considered, with the aim of grouping risk factors into homogeneous clusters as perceived by management and stakeholders (Figure 1).

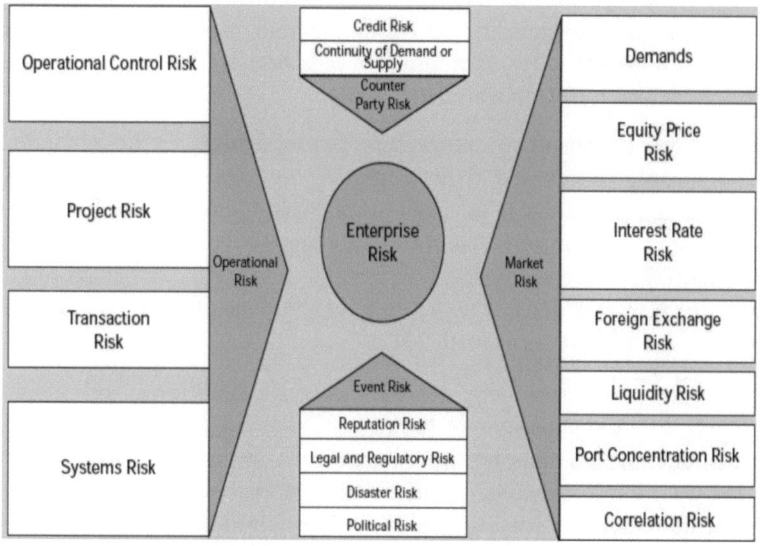

Fig. 1. Example of Enterprise Risk Classification (source: Clarke C. J. & Varma S., 1999)

Referring to most frequently adopted risk taxonomies (e.g. Tah & Carr 2001; Chapman, 2006), the basic operational risk categories can be identified as follows:

- Technology Risk: potential events in which the risk source is the technology implemented (i.e. poor performance of plants/equipments; failure in selecting a new technology, etc);
- Supply chain risk: potential events related to the procurement, expediting, inspection and logistic activities;
- Project risk: potential events affecting time, costs and quality objectives within project boundaries;
- Environmental risk: potential natural events impacting the area where the system/plant is located;
- Occupational risk: potential events affecting the employees health and safety;
- Information risk: potential events affecting critical requirements of information flows within the system/plant;
- Organizational risk: potential events related to lack of coordination, unclear task/objectives assignment, conflict or high turnover rate among the organization;
- Management risk: potential events caused by inadequate management processes or decisions. In these respect the complexity of the organization is the key driver of management risks;
- Facility and asset risks: potential events in which facilities or company assets are involved (e.g. fire).

2.2 Causal chains and influencing factors in Operational Risk Management

Despite their practical usefulness in allocating risk management responsibilities and simplifying risk reporting, operational risk classifications are largely inadequate to support the optimisation of risk control options, mainly in case of complex relationships among risk factors (e.g., interdependencies and escalation dynamics) or trade-offs between alternative lines of action . Indeed, operational risks are generated or influenced by a large spectrum of technology-, human- and organisational-related factors, that may dynamically combine together in several different ways, through complex and soft relationships that cannot be reduced to simple deterministic cause-effect chains.

Several examples can be raised to clarify this distinguishing nature of operational risks. Globalisation of supply chains and their increasing interconnectedness due to global and highly differentiated companies is an issue of increasing relevance that can be properly tackled only through more complex risk modelling approaches (Mittnik, S. & Starobinskaya, 2010). Similar requirements are needed when the relationships between global supply networks and critical infrastructures (electricity, gas, transportation, telecommunication, ...) are taken into consideration (Ferrari et al. 2011).

Also in project-based operations - e.g. aviation, power generation or oil & gas industries - traditional project risk management techniques (Chapman & Ward, 2003) are no longer sufficient to manage all the risks brought by modern large engineering projects. Indeed, interactions between project teams, company functions, business units and long term programmes create a network of interdependencies where a specific risk raising from a single project may create cascading effects climbing up at higher organisational levels, causing larger consequences than the one estimated at the project level (Silvestri et al. 2011).

In the last couple of decades common awareness on the increasing importance of human factors and organisational culture in shaping operational risks has also strengthen. Examples can be found in the analysis of the influence that safety culture may have on the occurrence of at risk behaviours and on injury rate in workplaces (De Ambroggi et al., 2008; Zhou et al., 2008), or in the increasing number of models proposed in literature to integrate human and organisational risk factors in Quantitative Risk Analysis (QRA) (Mohaghegh, Kazemi & Mosleh, 2009; Mohaghegh & Mosleh, 2009, Trucco et al. 2008a).

For all that, it comes clear that the effectiveness of Operational Risk Management practices can be improved only by providing the risk identification and risk analysis phases with enhanced risk modelling capabilities, able to take into account all the relevant contributing factors and mutual influences, from the root causes to the final effects. These emerging needs have to face two different but interrelated issues:

- the chronic lack of data and information on past events increases the importance of identifying and adopting proper methods to elicit experts' judgements and to assess epistemic uncertainties;
- the availability of different advanced risk modelling techniques - such as Bayesian Networks, System Dynamics (Sterman, 2000), Stochastic Petri Nets(Marsan et al. 1995), Fuzzy Cognitive Maps (Kosko, 1986) - foster the need of identifying clear driving criteria in the selection of the most appropriate one, under different risk management problems and application domains.

In the following sections of the chapter we offer a systematic review of the most interesting and relevant applications of Bayesian Networks and Bayesian Belief Networks to different problems in the area of Operational Risk Management. This critical overview is then used to identify and discuss some methodological issues and requirements for the correct adoption of BN in Operational Risk Analysis.

3. Modelling operational risks with BN: Critical review of the state of the art

"Probability theory is nothing but common sense reduced to calculation" Laplace, 1819.

The main issue in modelling operational risks has to do with the understanding of the functioning of a complex system. It requires the application of inductive logic for each one of the possible way in which a system operates to reach its objectives. Then it is the comparison between the hypothesis formulated in the functional analysis and the observations possible on the way the system actually function that can lead to an evolution of the knowledge regarding the system itself. This knowledge is the only credible base for the understanding and therefore a correct modelling of the system under analysis (Galvagni, 2011).

Therefore, the first feature that should be evaluated in a risk model is the functional analysis form which the modelling process stems.

The use of BBNs in modelling operational risk provides a specific advantage in respect to many other modelling approaches since a BBN is to be structured as a knowledge representation of the problem domain, explicitly including the probabilistic dependence between the main elements of the model and their causal relationship, therefore explicating the analyst's understanding of the problem. This is a key feature for validating the behaviour of the model and its accuracy in reporting to third parties the reality under analysis (Friiis-Hansen, 2000).

Furthermore, another issue that appears to be in common with all projects regarding the assessment of risks embedded in complex systems lays in the lack of consistent data. Example of this are for instance risk assessment studies on industrial plants willing to take into proper account human and organizational factors, where many analysts lament the lack of an adequate dataset for the quantification of the error mechanisms as well as for the contextual and organizational conditions affecting human performance (Straeter, 2004 and Fragola, 2000). Aside from this specific example in many operational domains the main issue regarding the assessment of safety and reliability of a system has to do with the scarce availability of data for the main causation factors to be taken into account. When data availability is a considerable issue the use of methods such that of Event Trees and Fault Trees would not be advisable for helping the analyst in the difficult issue of data gathering, especially because some of the data would be collected through the use of experts' judgments (Hensen, 2004). A more suitable method for implementing the main structure of safety assessment, as far as the causation factors for the accidental scenarios are concerned, is represented by the use of Bayesian Belief Networks (BBN). BBN are in fact better suited for representing uncertain knowledge. Further, since BBN approach stems from conditional independence assumptions and strongly relies on graphical representations, it makes it easy to display how the relationship among the variables and therefore the underlying data structure works. In addition, the outcome of compiling a model is the marginal probability distributions of all variables in the domain. Modelling local dependencies in facts amounts to specification of the probabilistic dependence of one variable on other variables. Therefore, even when the marginal distribution of the dependent variable is not known beforehand, it will be provided as a result of the assumptions being made on the causal relationships once the network has been compiled.

The main feature in this respect of BBN is that they allow easy inference based on observed evidence, even when the evidence to be observed is scarce. In fact, if one of the variables in the domain is observed then the probability distributions of the remaining variables in the model are easily updated accordingly. So, if the probabilities of a generic BBN are updateable, given a set of evidences collected from the field, a BBN model of organisational factors involved in accident scenarios might be validated over time, for instance, exploiting information contained in accident/incident reporting systems.

Specific examples where the pros and cons of using BBNs have already been explored are the followings:

- Integration of human and organisational risk factors in system safety engineering;
- Safety culture analysis and assessment;
- Project Risk Management;
- Operational Risk Management (ORM);
- Integration between Enterprise Risk Management (ERM) and ORM.

3.1 BBN and Human and Organizational Factors (HOF) in Probabilistic Risk Analysis (PRA)

BBN are becoming more and more widely used in the current generation of Probabilistic Risk Analysis (PRA), to try and support an explicit representation of the possible impacts of organization and management processes on the safety performance of equipment and personnel (Trucco et al. 2008a).

In the Bayesian statistical framework, a fully quantified BBN represents the prior knowledge for the analyst. However, as already pointed out, the model can be updated using observations (sets evidence) about certain nodes and verifying the impact on the remaining nodes in the network. By setting evidence, an analyst is proving the model with new information (e.g., recent incident events) about the state of the system. And this information can be propagated through the network to produce updated probabilities for all nodes in the model. These resulting probabilities combine both prior information and new evidence. BBNs have been recently used in traditional Probabilistic Risk Analysis by linking BBN nodes to other risk models using the so called Hybrid Causal Logic methodology (Groth et al., 2010; Wang, 2007), which links BBNs to Event Trees and Fault Trees. The use of HCL enables to include soft causal factors, such as human error in more deterministic models, which were more traditionally used for hardware systems.

Furthermore, current HRA methods often ignore the interdependencies and causal relationships among various Performance Shaping Factors (PSFs). While only recently BBNs have been proposed as a way of assessing the interactions among PSFs and the failure modes they are suppose to influence (Fig. 2; Leva et al., 2006; Groth, 2009).

The model used by Leva et al. (2006) for assessing human performance in a solo watch situation for a ship on possible collision courses takes into account the main elements affecting human performance considering features of the ship that are also observable during a normal training session with the use of a bridge simulator. Thus the time to detect a ship, the time used for planning an action, the probability of taking the wrong decision the probability of performing the wrong execution of a manoeuvre (even if the right plan has been made) and the needed time for manoeuvring the ship have been considered as the primary elements of the operator performance in the model. As most of the Human Reliability Models also the data used for the current example mostly rely on experts' Judgments. However the model was built so as to collect and make use of real observational data (collectable from observations, as, for instance, training sessions) this should be the final test of any model: the verification coming from experience.

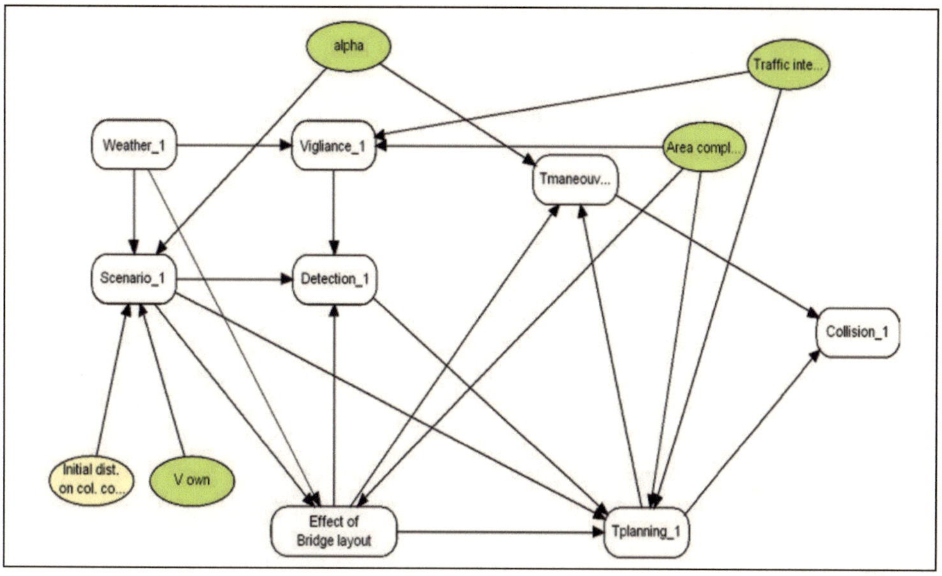

Fig. 2. Example of a BN used for the assessment of an operator not reacting in time in a ship collision scenario. The elements in white with a rectangular shape are object (sub-networks) while the nodes are input nodes to be inserted by the end-user. (Source: Leva et al. 2006).

However as pointed out by Mohaghegh and Mosleh (2009) there are a number of technical challenges in developing a predictive model of organizational safety performance most of which have to do with "the absence of a comprehensive theory, or at least a set of principles and modelling guidelines rooted in theory and empirical studies" as the major cause of current lacking of an adequate basis to validate these models. Yet as already pointed out if the probabilities of a generic BBN are updateable given a set of evidences collected from the field, a BBN model of organisational factors involved in accident scenarios for instance might be validated over time exploiting information contained in accident/incident reporting systems. So why is it that this empirical validation is often missing from the literature?

Looking at the characteristics of several HOF models proposed in literature, it seems to us that, in general it is their increasing complexity that mainly impedes to clearly justify modelling solutions, to assure consistency, replicability, and eventually the possibility to sue observation data for validation purposes. This issue might be particularly critical when multiformalism is adopted: limitations posed by the integration of different sub-models often weaken the quality and the detailed specification of single parts of the model and BBNs are therefore often mixed with other modelling formalisms used to model interconnected parts of a final PRA contributing model (e.g. operator model, system model, etc.) (Trucco & Leva, 2010). So the attempt to incorporate an even broader spectrum of soft factors - such as safety culture, climate, management commitment to safety, etc. - requires to develop complex but ambiguous HOF models where the main weakness is the measures of hardly measureable factors, and results in what Dougherty calls an "often obfuscating numerology"(1990).

3.2 The use of BBN to assess safety culture

The validity of BBNs in supporting the modelling of safety culture and the evaluation of potential strategies for safety improvement has been demonstrate by Zhou et al. (2008) when they proposed a Bayesian Network (BN) based model aimed at establishing a probabilistic relational network among causal factors, including safety climate factors and personal experience that were thought to have an influence on human behaviour pertinent to construction safety. Zhou et al. (2008) study used the data coming from a survey involving more than 4700 employees at a large construction firm to collect the data to feed the network. The BBN was built around the categories used in the survey based on theoretical models previously developed about the factors affecting safety climate. The results of the study, and consequently the factors to be considered, were revised on the bases of the results of the factorial analysis. The scope of the BBN developed was to support the diagnosis of the state of a safety climate, the diagnostic of main issues and consequently the identification of potential strategies for safety improvement. The use of BBNs for representing, analysing and improving the actual anatomy of company's safety culture and its impact on the expected probability of safe behaviours performed by workers was also

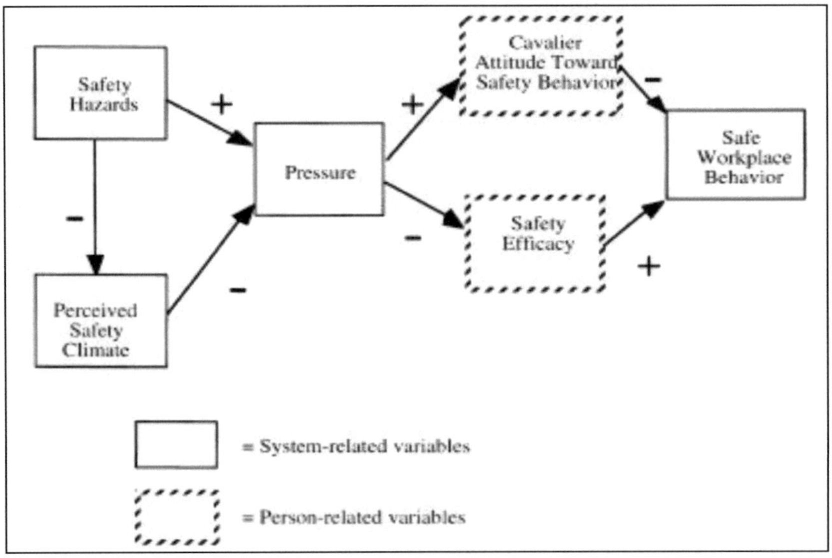

Fig. 3. Preliminary socio-technical model predicting safe work behaviour (Source: Brown et al., 2000).

used in successive studies (e.g., Trucco et al., 2008b), in some of them the results of the survey were used to find out the Bayesian structure underlying the relationships among socio-technical factors. This is possible through an algorithm called K2 (Cooper & Herskovitz, 1992). The BBNs resulting from the use of the algorithm are then often reviewed by the experts to direct the arcs in the direction that makes more sense in terms of cause-effect relationships (e.g. it is apparent, for example, that the "age of the worker" affects the safety climate and not the reverse) and an underlying theoretical model can also be used as a guiding principle (Figure 3).

Trucco et al. (2008b) applied the proposed methodology to identify and analyse the effectiveness of different organizational and behaviour-based measures for improving occupational safety in a leading tractor manufacturer. The BBN representation of the safety culture structure in the manufacturing area is reported in Figure 4.

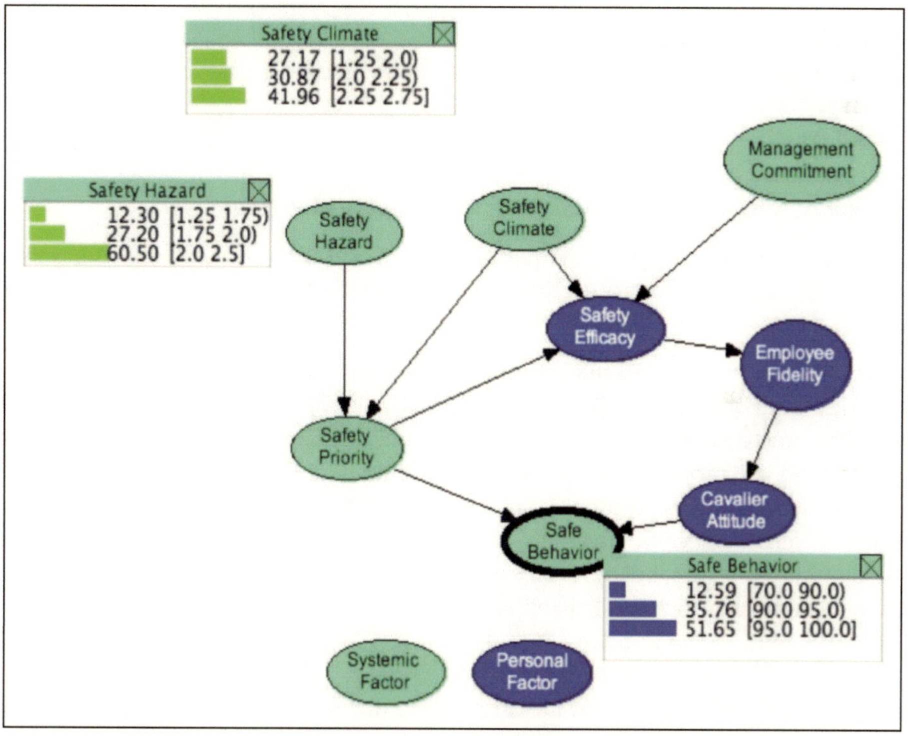

Fig. 4. Example of sensitivity analysis on some safety culture variables in the manufacturing area of a large truck manufacturer (Trucco et al., 2008b).

Considering the current setting of systemic factors as assessed by employees (e.g. 27.2% probability of having poor safety climate, 30.8%, for good safety climate and 42% for optimal safety climate), the rate of safe work behaviours was estimated about 93.6%. Even though this value may seem high (6 unsafe behaviours out of 100), the high value of the severity index of incidents occurred at workers operating in manufacturing area suggests the need for an improvement of compliance with safe behaviours. Table 1 reports a summary of the expected effectiveness of alternative strategies to improve the rate of safe work behaviours.

Strategy	Safety hazards (%)	Safety Climate (%)	Management commitment (%)	**Safe Work Behaviour (%)**	**Unsafe Work Behaviour (%)**
Single factor optimisation (simple strategy)	-27,2			**0,8**	**-11,9**
		16,3		**0,0**	**-0,2**
			13,7	**0,2**	**-3,1**
10% single factor improvement (simple strategy)	-10			**0,3**	**-4,4**
		10		**0,0**	**-0,1**
			10	**0,2**	**-2,3**
Multiple factors optimisation (complex strategy)	-27,2	16,3		**1,5**	**-21,3**
	-27,2		13,7	**1,0**	**-14,7**
		16,3	13,7	**0,3**	**-3,9**
	27,2	16,3	13,7	**1,8**	**-25,9**

Table 1. Summary of the expected effectiveness of strategies to improve safe work behaviours in the manufacturing area (values are in percentage; negative values means decreases) (Trucco et al., 2008b).

3.3 The use of BBN and risk assessment in project management

BBN have been recently applied to quantify the probability of risks affecting success of projects like for instance the probabilities of significant delays (Luu et al., 2008; Wang et al., 2009).

BBNs have in fact been usefully deployed in the area of decision support under uncertainties (Bouissou et al., 1997; Ziv & Richardson, 1997). There are many uncertainties in development processes for products of processes like the uncertainties in estimating project completion time, the project needs for supply the quality of the output etc. From experience or from the literature it is to identify the main factors related to delays in projects. The literature can also be specific about the domain the project risk factors relate to, such as construction industry (Assaf et al., 2006), or software development projects (Fan & Yu, 2004) and Hi-Tech industry (Raz & Michael, 2001). However some factors are also in common across the different domains: delay antecedents for instance can be factors caused by clients, contractors, consultants, and designers, or to the main inputs (Materials-, workforce-, and equipment-related factors are input factors); environment-related factors (exogenous factors such as difficult meteorological conditions, changes in government regulations and laws, traffic control etc.); Project-related factors are factors deriving from the project characteristics and the way the process is designed to deliver the desired outcome.

The usefulness of a BBN based approach in assessing the projects associated risks and the likely outcomes can be summarised as follow:

- Help to perform continuous risk management using data collected as the project develop to provide a feedback loop to detect and adjust problematic situations , as shown in Figure 5 (Lee et al., 2009).
- As already said, the BBNs model can take into account the main uncertainties and provides probabilistic estimates for them. Whenever new evidence is available in the

monitoring loop, the new data can be plugged in the related BBNs model to recalculate and update previous estimates.

- Moreover a model developed for one project may help identifying and evaluating the relative importance of the significant factors contributing to delay cost overruns in general on the basis of the actual collection of statistical evidence (Luu et al., 2009). This in turns can also help modifying the model itself as belief networks also allows variables to be added or removed without significantly affecting the remainder of the network because modifications to the network may be isolated (McCabe et al., 1998)

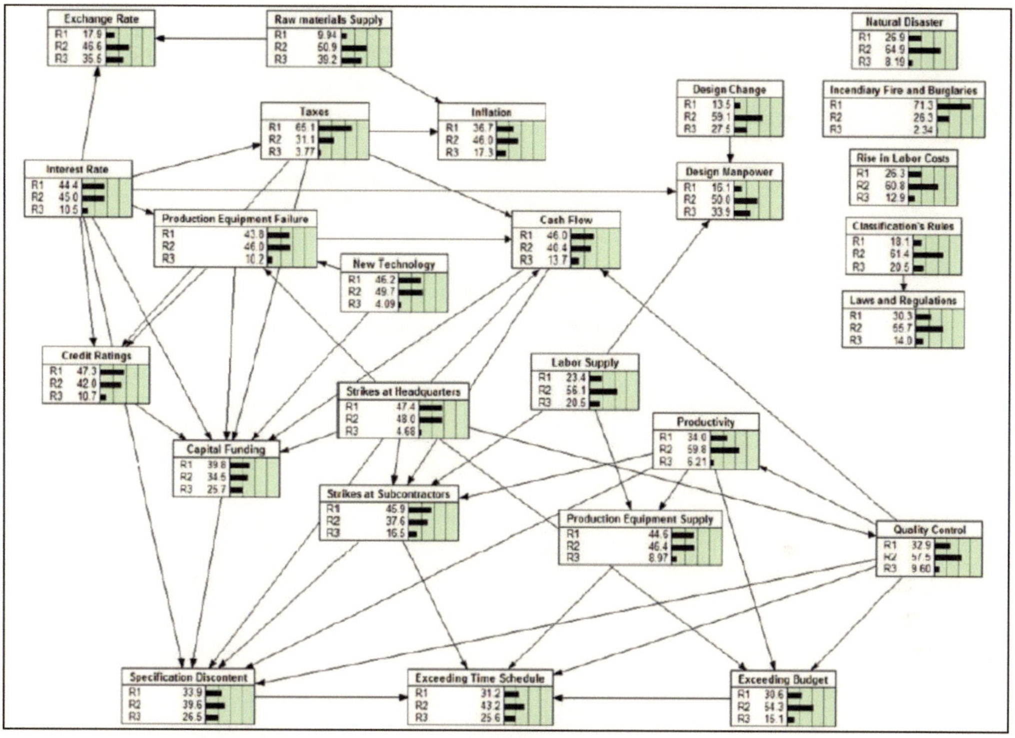

Fig. 5. Example of A BBN used for predicting project issues in shipbuilding (source: Lee et al., 2009).

3.4 From assessing risks in project management to operations risk management: advantages of BBN approaches

Operational risks have also been defined as risks of human origin that, unlike financial risks that can be handled in a financial manner (e.g. insurances, savings, derivatives), require a more "managerial approach"(Fragniere et al., 2010).

The recent developments in the quantification of Operational Risk has, to a significant extent, been determined by changes in the supervisory regimes for financial institutions. These changes have increased the level of supervisory scrutiny on Operational Risks (OR) and how it is managed by relevant firms has been deeply affected by the high-profiled corporate failures in recent decades. This has determined the development of Operational Risks models as a means to demonstrate good management and financial strength (Cowell et al., 2007). Even in this domain Bayesian Networks offer a way to combine both qualitative and quantitative data and also to meet the requirements of the regulators for measuring OR. As pointed out by Conalba and Giudici (2004) the use of Bayesian networks for operational risk management allows to integrate, via the Bayes' theorem, different sources of information coming from loss data collection, self assessment, industry loss data and opinion of risk managers, to give a unified knowledge. This capacity in turns facilitate the managing of OR (i.e., identification, assessment, monitoring and control/mitigation) and justify decision taken on a more transparent ground, combining the use of retrospective historical data with prospective expectations and opinions so as to evaluate also the Influence of "causal" factors (Cornalba & Giudici, 2004). Summarising, the usage of BNs in modelling OR loss distribution, can have significant benefits for supporting decisions, particularly in capital allocation. Stress and scenario testing are also possible in BBNs allowing the drafting of an early warning system (Figure 6; Yoon, 2003).

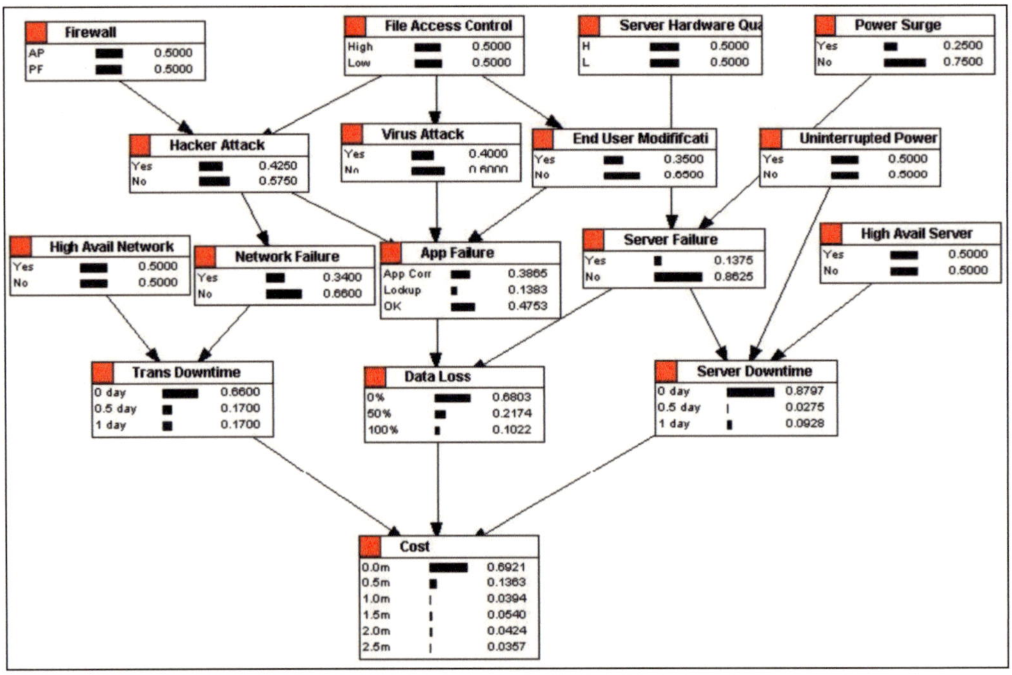

Fig. 6. Example of the prior distribution assigned to a BBN used for predicting costs derived from operational risks (source: Yoon, 2003).

3.5 The use of BBNs to support Enterprise Risk Management

Enterprise risk is normally defined as the possibility that something with an impact on the company objectives happens, and it can be measured in terms of combination of probability of an event (frequency) and of its consequence (impact).

Enterprise risk assessment is a keystone of Enterprise Risk Management (ERM) therefore it is vital for the assessment to be as much as possible grounded on trustworthy assumptions.

Bonafede and Giudici (2006) have reported that to estimate the frequency and the impact distributions historical data as well as expert opinions are typically used. Then such distributions are combined to get the loss distribution. In the case of enterprise risk assessment the considered risks can be strategic, operational, legal and political and they are normally difficult to quantify. As for many other domains also in this case it is often easier to gather data from experts' opinions. In this context Bayesian Network are a useful tool to integrate historical data with qualitative or quantitative estimates coming from experts. Example of applications are the use of BBN to examine the risk related to production or distribution or certain products (Pai et al., 2003) or the ones associated to specific decisions in the management of a business like the risk involved in the choice of a supplier or in outsourcing a certain service/activity. The example provided by Lockami and McCormack (2010) for instance is a BBN model that examines the probability of a supplier's revenue impact on a company based upon the supplier's associated network, operational, and external risks. Network, operational, and external risks were determined based upon the a priori probabilities for risk events which directly influence them.

Figure 7 reports the network they developed in their study. The nodes named with numbers represent the set of considered potential influencing factors: misalignment of interest (1); supplier financial stress (2); supplier leadership change (3); tier stoppage (4); supplier network misalignment (5); quality problems (6); delivery problems (7); service problems (8); supplier HR problems (9); supplier locked (10); merger/divestiture (11); disasters (12). The model was found useful for supporting outsourcing decisions, develop risk profiles for suppliers so as to analyse current and future outsourcing relationships. However, as noted by the authors, the most important potential limitation to the use of this methodology to assess risks in supply networks is the ability to provide accurate information regarding external risks as reflected in the 12 risk events outlined in the model.

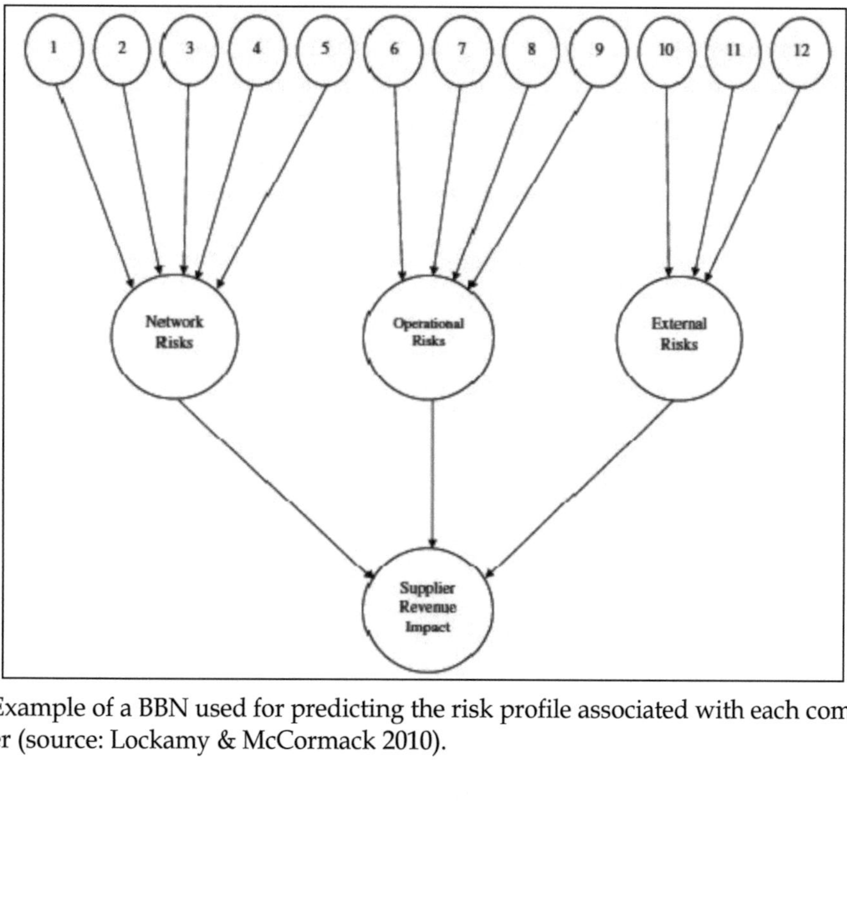

Fig. 7. Example of a BBN used for predicting the risk profile associated with each company Supplier (source: Lockamy & McCormack 2010).

In this domain BBN are often used as influence diagrams. An influence diagram is a Bayesian Network used for the scope of solving decision problems and it presents some special features. In an influence diagram two additional types of nodes are included in the network, namely decision nodes (rectangular shaped) and utility nodes (diamond shaped). A decision node defines the action alternatives that the user is considering. Preceding nodes on decision nodes define information available at the time of decisions. Decision nodes may have multiple children, and thus dependent on the choice of action alternative the decision node changes the state of the world. On the other hand utility nodes have no children but are conditioned on probabilistic and/or decision nodes. The utility nodes hold tables of utility for all possible configurations of the outcomes of the parent nodes. The rational basis for decision-making is established by computation of the expected utility (EU) of each of the action alternatives. Being an influence diagram a modified Bayesian Network, evidence can be inserted into the model. Propagating this evidence can give updated expected utilities for all decision variables. Hence as Hensen (2004) points out "the influence diagram serves as a dynamic decision model always showing the optimal strategy, possibly conditional on a set of observations. The optimal plan initially suggested may therefore be altered, as more data becomes available. Moreover, the expected utilities of the non-optimal choices are always available allowing a quantitative comparison of the action alternatives. However it should be noted, that when a Bayesian Network is combined with decision nodes it is essential that the Bayesian Network is modelled as a causal model since in an influence diagram the flow of information can only follow the causal link". However the modelling domains of enterprise risk assessment are often so complex that it is intrinsically difficult to establish clear causal relationship among all the variables at play.

4. Methodological issues and requirements for BBN applications in risk analysis

Looking at the BBN applications presented in the previous sections it is clear that compared to other analysis tools, they offer several capabilities to a risk analyst that has to face different types of risk factors and mechanisms involved in complex socio-technical system. Moreover, when needed, BBN risk models can be easily reduced to more traditional risk analyses as in the case of structural reliability problems studied through the so-called max-propagation (Friis-Hensen, 2004); the algorithm returns the most probable configuration of the network given the occurrence of a specified event (hard evidence). When some critical failure events are taken into consideration, the max-propagation algorithm can be used to identify the most probable configuration of the network (system's risk factors) that leads to the occurrence of a specific critical event. If the nodes of the BN are binary variables, the max-propagation directly gives the most dominant cut-set as well as the application of the Fault Tree Analysis (FTA) of the same system.

In the previous paragraphs however we have been mainly focusing on the benefits of using BBN to model operational risk in various domains. Nevertheless, also BBNs have limitations and shortcomings. The adoption of a coherent modelling approach is thus a key element for assuring the relevance, the accuracy and the reproducibility of the risk model. In this regard some issues are worth to be considered:

- Before starting to define the topology of the BN model it is very important to fully understand the structure and the dynamics of the system first and the scope of the analysis as well. This statement may be perceived as obvious, but nowadays Bayesian Networks are often built through very intuitive graphical software and it is therefore very easy to get carried away by the graphical modelling, that at the end may be incoherent and misleading;

- The model can get highly complex very quickly with many nodes and relationships to be specified - this is especially true when the nodes have many parents. Indeed, the size of a CPT grows exponentially with the number of the parents; for example, a node with five parent variables, defined with only three states, requires the specification of 243 entries for each one of its states. In such a case, there can be too many conditional probabilities to specify - if the maximum likelihood method of prior elicitation is used, significant volume of data might actually be required, thus reducing one of the main advantages of using Bayesian methods;
- BBNs pose the problem of trustworthy exert opinion elicitation. Sometimes this would need the deployment of rigorous methodologies in prior elicitation through costly methods, such as the Delphi method which involves many rounds of questionnaires. Another issue can derive from the fact that sometimes the experts are not comfortable in eliciting frequencies (Yoon, 2003);
- Last but not least, BBNs generally require that the state space of nodes shall be countable and discrete; thus their application require the discretisation of random variables. Discretisation is not simple and when applied to variables continue in nature, sometimes brings to the definition of many categories and therefore many possible states. This is a downside of BBN strictly connected to the previous issue, i.e. the exponential growth of the number of states and thus of the dimension of CPTs. However, as Friis-Hansen (2004) points out, neither Fault Tree Analysis (FTA) nor Event Traa Analysis (ETA) offer any better alternatives.

Cowell et al. (1999) in their book provide useful guidelines on how to deal with these methodological issues.

5. Conclusions

In the realm of risk assessment of modern complex socio-technical systems, as already mentioned, it is of paramount importance the identification and understanding of all the causal chains leading to disruptions or even destruction of the system. Several internal and external factors of different nature - human, organisational, natural, sociological, political - may influence or modulate these cause-effect mechanisms and must be taken into proper consideration. It requires the application of inductive logic for each one of the possible way a system operates to reach its objectives. This knowledge is the only credible base for the understanding and therefore a correct modelling of the system under analysis (Galvagni, 2011). The main advantage provided by the use of BNs in modelling operational risks is that the model itself can be structured as a knowledge representation of the problem, explicitly including the probabilistic dependence between the main elements of the risk model and their causal relationships, therefore explicating the level of understanding achieved by the analyst. This is a key feature for validating the accuracy of the risk model and its reliability in reporting to third parties (Friis-Hansen, 2000).

Furthermore as seen for the case of applications to project risk assessment, BBNs are able to provide a way of comparing the cost of the action to its risk mitigating effect. Similarly, in applications regarding OR it is of great importance to carefully evaluate whether the expected risk reduction for the considered initiative is worth its estimated cost. In the end, being able to provide a more transparent and rational ground to decision makers is really key. Moan (2000) clearly illustrates the benefit of rational evaluations in risk management.

However, as briefly discussed in Section 4, the specification of the structure of a BBN is often subject to debate because based on expert assumptions and/or on theoretical modelling of the reality under analysis, that have not been subject to the test of operational experience. For this reason the tendency is to deploy the BBNs capability of using real data for structural learning - i.e. letting the data speak for itself not just with regards the probability distributions of the variables but even the very structure itself (Yoon, 2003). This is currently a promising research area.

6. References

Abbott A. (1988). *The systems of profession*. Chicago University Press, Chicago, USA.

Assaf, SA & Al-Hejji, S. (2006). Causes of delay in large construction projects. *Int J Project Manage*, Vol.24, No.4, pp. 349–57.

BCBS (2001). *Working paper on the regulatory treatment of operational risk*. BCBS Working Papers No 8. Basel Committee on Banking Supervision, Bank for International Settlements. Available from: http://www.bis.org/publ/bcbs_wp8.htm

Beroggi, G.E.G. & Wallace, W.A. (1994). A Decision Logic for Operational Risk Management. *Computational Organization Theory* (Edited by Carley, K. & Prietula, M.), LawrenceErlbaum Associates, Hillsdale, USA.

Bonafede, C.E. & Giudici, P. (2006). *Construction of a bayesian network for a project of enterprise risk management*. Technical report, University of Pavia, Italy.

Bouissou, M., et al. (1997). Assessment of a safety-critical system including software: a Bayesian Belief Network for evidence. *IEEE Proceedings of Annual Reliability and Maintainability Symposium*, pp. 142–150.

Brown, K. A., Willis, A. G. & Prussia G. E. 2000. Predicting safe employee behaviour in the steel industry: Development and test of sociotechnical model. Journal of Operations Management 18(4): 445-465.

Chapman (2006). *Simple tools and techniques for Enterprise Risk Management*. The Wiley Finance Series, John Wiley and Sons, Chichester, England, UK.

Chapman, C. B. & Ward, S.C. (2003). *Project Risk Management: Processes, Techniques, and Insights*. John Wyley & Sons, Chichester, England, UK.

Clarke C. J. & Varma S. (1999). Strategic Risk Management: the New Competitive Edge. *Long Range Planning*, Vol. 32, pp. 414-424.

Cooper, G. F. & Herskovitz, E. A. (1992). Bayesian method for the introduction of probabilistic networks from data. *Machine learning*, Vol.9, pp. 309-347.

Cornalba, C. & Giudici, P. (2004). Statistical models for operational risk management *Physica A: Statistical Mechanics and its Applications*, Vol.338, Nos.1-2, pp. 166-172.

Cowell, R.G., Dawid, A. P., Lauritzen, S.L. & Spiegelhalter, D.J. (1999). *Probabilistic Networks and Expert Systems*. Springer, New York, USA.

Cowell, R. G., Verrall, R. J. & Yoon, Y. K. (2007). Modeling Operational Risk With Bayesian Networks. *Journal of Risk and Insurance*, Vol.74, pp. 795–827.

Fan, C. & Yu, Y.-., 2004. BBN-based software project risk management. Journal of Systems and Software, 73(2), pp. 193-203.

Ferrari, M, Schupp, B.A., Ward, D., Nordvik, J.P. & Trucco, P (2011). Assessing Supply Chain dependency on Critical Infrastructures using Fuzzy Cognitive Maps. *International Journal of Risk Assessment and Management* (Special Issue on: "Risk Analysis of Critical Infrastructures"), Vol. 15, Nos. 2/3, pp. 149-170.

Fragnière E., Gondzio J. & Yang X. (2010). Operations risk management by optimally planning the qualified workforce capacity. *European Journal of Operational Research*, Vol.202, No.2, pp. 518-527.

Friis-Hansen, A. (2000). *Bayesian Networks as a decision Support tool in Marine Applications.* PhD Thesis. Department of Naval Architecture and Offshore Engineering, Technical University of Denmark December.

Friis-Hansen, P. (2004). Structuring of complex systems using Bayesian networks. *Proceeding of Two Part Workshop at DTU*, August 23-25, pp. 110-133.

Groth K., A (2009). *Data-informed model of performance shaping factors for use in human reliability analysis*, Ph.D. dissertation, University of Maryland, College Park, MD.

Kaplan, R.S., Anette, M., Simons, R., Tufano, P. & Hofmann, M. (2010). Managing Risk in the new world. *Harvard Business Review*, 87/10.

Kosko, B. (1986). Fuzzy cognitive maps. *International Journal Man-Machine Studies*, Vol.24, pp.65-75.

Lee E., Yongtae P. & Shin J.G. (2009). Large engineering project risk management using a Bayesian belief network. *Expert Systems with Applications*, Vol.36, pp. 5880–5887

Leva, M.C., Hansen, P.F., Sonne Ravn, E. & Lepsøe, A. (2006). SAFEDOR: a practical approach to model the action of an officer of the watch in collision scenarios. *Proceedings of ESREL Conference*, Estoril Portugal, Taylor & Francis Group.

Lockamy A. & McCormack K. (2010). Analysing risks in supply networks to facilitate outsourcing decisions, *International Journal of Production Research*, Vol.48, No.2, pp. 593-611.

Luu, V.T., Kim, S., Tuan, N.V. & Ogunlana, S.O. (2009). Quantifying schedule risk in construction projects using Bayesian belief networks. *International Journal of Project Management*, Vol.27, No.1, pp. 39-50.

Marsan, M. A., Balbo, G., Conte, G., Donatelli, S. & Franceschinis G. (1995). *Modelling with Generalized Stochastic Petri Nets*. Wiley Series in Parallel Computing, John Wiley and Sons, Chichester, England, UK.

McCabe, B, AbouRizk, SM & Goebel R. (1998). Belief networks for construction performance diagnostics. *J Comput Civil Eng ASCE*. Vol.12, No.2, pp.93–100.

Mittnik, S. & Starobinskaya, I. (2010). Modeling dependencies in operational risk with hybrid Bayesian Networks. *Methodology and Computing in Applied Probability*, Vol.12, No.3, pp. 379-390.

Mohaghegh, Z., Kazemi, R. & Mosleh, A. (2009). Incorporating organizational factors into Probabilistic Risk Assessment (PRA) of complex socio-technical systems: A hybrid technique formalization. *Reliability Engineering and System Safety*, Vol.94, No.5, pp. 1000-1018.

Mohaghegh, Z. & Mosleh, A. (2009). Measurement techniques for organizational safety causal models: Characterization and suggestions for enhancements. *Safety Science*, Vol.47, No.10, pp. 1398-1409.

Pai, R. R., Kallepalli, V. R., Caudill, R. J. & Zhou, M. C. (2003). Methods toward supply chain risk analysis," in *Proc. of 2003 IEEE Int. Conf. on Systems, Man, and Cybernetics*, 4560-4565, Washington D.C.

Power, M (1993). *Organized uncertainty*, Oxford University Press, Oxford, UK.

Raz T. & Michael, E. (2001). Use and benefits of tools for project risk management International *Journal of Project Management,* Vol.19, No.1, pp. 9-17.

Samad-Khan A. (2008). Modern Operational Risk management. *Emphasis*, 8/2, Available from http://www.towerswatson.com.

Silvestri, A., Cagno, E. & Trucco, P. (2009). On the anatomy of Operational Risk. In *Proceedings of IEEE International Conference on Industrial Engineering and Engineering Management*, 8-11 December, Hong Kong (China).

Silvestri (2010). *New framework and modeling approach to Enterprise Risk Management in E&C organizations, with application in the Oil & Gas.* PhD Thesis. School of Management, Politecnico di Milano, Milan, Italy.

Silvestri, A., Arena, M., Cagno, E., Trucco, P. & Azzone, G. (2011). Enterprise Risk Management from Theory to Practice. The Role of Dynamic Capabilities Approach - the "Spring model", in *Quantitative Financial Risk Management*, Wu, D.D. (Ed.), Springer Verlag, New York, USA.

Sterman, J.D. (2000). *Business Dynamics: Systems thinking and modeling for a complex world.* McGraw Hill. ISBN 0-07-231135-5.

Tah & Carr (2001). Towards a framework for project risk knowledge management in the construction supply chain. *Advances in engineering software*, Vol. 32, pp. 835-846.

Taleb N.N., Goldstein D.G. & Spitznagel M.W., (2009). The six mistakes executives make in risk management. *Harvard Business Review*, 87/10.

Trucco, P., Silvestri, A. & Cagno, E. (2010). Understanding and Modeling Operational Risks: A Multi Context Application. *In Proceedings of PSAM10 Conference*, 6-11 June, Seattle (WA), USA.

Trucco, P., Cagno, E., Ruggeri, F. & Grande, O. (2008a). A Bayesian Belief Network modelling of organisational factors in risk analysis: A case study in maritime transportation. *Reliability Engineering and System Safety*, Vol.93, No.6, pp. 845-856.

Trucco, P., De Ambroggi, M. & Grande, O. (2008b). A Bayesian Belief Network Model for Assessing the Anatomy and Effectiveness of Safety Culture: a Case Study on Occupational Safety in the Automotive Industry. In Martorell et al. (eds): *Safety, Reliability and Risk Analysis: Theory, Methods and Applications*, CRC Press, Taylor & Francis Group, London, UK.

Yoon Y. K. (2003). *Modelling Operational Risk In Financial Institutions Using Bayesian Networks.* Dissertation submitted for the Master of Science Faculty of Actuarial Science and Statistics Cass Business School City of London.

Zhou, Q., Fang, D. & Wang, X. (2008). A method to identify strategies for the improvement of human safety behavior by considering safety climate and personal experience. *Safety Science*, Vol.46, No.10, pp. 1406-1419.

Ziv, H. & Richardson, D.J. (1997). Constructing Bayesian-network models of software testing and maintenance uncertainties. In: *Proceedings of International Conference on Software Maintenance*, pp. 100–109.